Praise for
The Worm Farmer's Handbook

"In my 24 years in the worm business, I have read and sold many books about worms. *The Worm Farmer's Handbook* is the most comprehensive and detailed collection of information on vermicomposting I have encountered. Rhonda Sherman provides clear instructions and guidance on everything a beginner needs to know to start up and successfully maintain a worm business, from type of beds and feedstocks to how to feed, harvest, package, ship, and much more. Experienced farmers will find plenty of excellent tips and ideas to improve their businesses, too. Well done, Rhonda!"

—**Mark Purser**, co-owner, The Worm Farm, Durham, California

"My very first business venture as a child was a worm farm, which failed horribly. Had I had Rhonda Sherman's mentorship, I can only imagine the worm empire I'd have today. With *The Worm Farmer's Handbook* as a guide, anyone can learn the science and art of raising worms."

—**Tony Hsieh**, CEO, Zappos; author of *Delivering Happiness*

"Rhonda Sherman masters the art of communicating the technical and biological minutia critical to successful worm farming in a personable and easy-to-understand manner. The case studies in *The Worm Farmer's Handbook* bring vermicomposting alive! This book is the next best thing to being on an actual worm farm tour. Kudos!"

—**Nora Goldstein**, editor, *BioCycle* magazine

"Rhonda Sherman's book has arrived on the scene at just the right time, as vermiculture technology is gaining momentum in research and industry. The book combines Sherman's hands-on experience with detailed descriptions of current practices and comprehensive documentation of research findings, making it a vital reference for worm farming practitioners and enthusiasts."

—**Norman Q. Arancon**, PhD, associate professor of horticulture,
University of Hawaii at Hilo

"*The Worm Farmer's Handbook* is a must-read for anyone interested in vermicomposting or vermiculture. With a candid voice, Rhonda Sherman explains the pitfalls and opportunities of worm farming based on her decades of experience. Chock full of tips for replication and lessons learned from model enterprises, the book's chapters span not only the globe but a plethora of systems and sizes."

—**Brenda Platt**, director, Composting for Community Project,
Institute for Local Self-Reliance

"With *The Worm Farmer's Handbook,* Rhonda Sherman delivers the first concise book of its kind: The A through Z of vermicomposting for commercial and community-scale enterprises."

—**Frank Franciosi**, executive director,
US Composting Council

"Rhonda Sherman sheds light on the mysterious world of worms and provides valuable insight into the worm composting industry. *The Worm Farmer's Handbook* is my new first stop for practical, trustworthy vermicomposting information."

—**James McSweeney**, author of
Community-Scale Composting Systems

"Rhonda Sherman's comprehensive new manual covers everything about creating a vermiculture business, from writing a business plan, marketing, and the care and feeding of worms to choosing a system that works for you. I wish her book had been available when I bought my worm farm 25 years ago! For over 20 years I have believed that vermiculture has the potential to revolutionize agriculture. *The Worm Farmer's Handbook* can help that revolution go forward, and I heartily recommend this book to anyone interested in vermiculture."

—**Jack Chambers**, president, TerraVesco

"Before reading this book, I made a mental note of topics that I felt would be included, such as history, how-to, underlying science, pitfalls and benefits, and economics. I was not disappointed—these topics and so much more are fully covered and written about in a very readable manner. This book will prove useful to academics and students at every educational level, practitioners of all types, and anyone with an enquiring mind. In an ever-growing field, *The Worm Farmer's Handbook* is set to be the must-have guide and is a very welcome addition to earthworm literature."

—**Kevin R. Butt**, PhD, director, Earthworm Research Group,
University of Central Lancashire

"With global topsoil depletion reaching alarming levels, this thorough, informed, detailed, and concise manual could not be more timely. Worm farming presents unprecedented hope and enterprising opportunities that are being adopted in every corner of the world. All food security activists, healthy soil advocates, farmers, and educators should read *The Worm Farmer's Handbook.* Thank you, Rhonda—this book is a highly useful gift to us all. The world needs more worm farmers!"

—**Anna de la Vega**, director,
The Urban Worm Community Interest Company

THE WORM FARMER'S HANDBOOK

Mid- to Large-Scale
Vermicomposting for Farms,
Businesses, Municipalities,
Schools, and Institutions

Rhonda Sherman

Chelsea Green Publishing
White River Junction, Vermont
London, UK

Project Manager: Sarah Kovach
Editor: Fern Marshall Bradley
Copy Editor: Laura Jorstad
Proofreader: Eileen M. Clawson
Indexer: Andrea Chesman
Designer: Melissa Jacobson
Page Composition: Abrah Griggs
Editorial Assistant: Nick Kaye

Printed in the United States of America.
First printing October, 2018.
10 9 8 7 6 5 4   22 23 24

Our Commitment to Green Publishing

Chelsea Green sees publishing as a tool for cultural change and ecological stewardship. We strive to align our book
manufacturing practices with our editorial mission and to reduce the impact of our business enterprise in the environment. We print our books and catalogs on chlorine-free recycled paper, using vegetable-based inks whenever possible.
This book may cost slightly more because it was printed on paper from responsibly managed forests, and we hope you'll
agree that it's worth it. *The Worm Farmer's Handbook* was printed on paper supplied by Versa Press that is certified by the
Forest Stewardship Council.

Library of Congress Cataloging-in-Publication Data

Title: The worm farmer's handbook : mid-to large-scale vermicomposting for farms, businesses,
   municipalities, schools, and Institutions / Rhonda Sherman.
Description: White River Junction, Vermont : Chelsea Green Publishing, [2018]
   | Includes bibliographical references and index.
Identifiers: LCCN 2018028054| ISBN 9781603587792 (pbk.) | ISBN 9781603587808 (ebook)
Subjects: LCSH: Earthworm culture. | Vermicomposting.
Classification: LCC SF597.E3 S45 2018 | DDC 639/.75—dc23
LC record available at https://lccn.loc.gov/2018028054

Chelsea Green Publishing
85 North Main Street, Suite 120
White River Junction, VT 05001
(802) 295-6300
www.chelseagreen.com

For Ian

Contents

Introduction

The ad says, "Make money selling worms." You may wonder, *Is that really possible?* It certainly is! In 2015 a segment of the television show *Blue Collar Millionaires* featured a worm farmer in Northern California because—you guessed it—he's a millionaire. That farmer has attended almost all of the North Carolina State Vermiculture Conferences I have organized over the years (2018 will be the 19th annual), and we have become friends. You'll learn about his worm farm in chapter 10 on page 217.

Tony Hsieh, the billionaire owner of Zappos (the online shoe and clothing company), ended the introduction in his best-selling book *Delivering Happiness* with the line, "My path began on a worm farm." To promote the book, Hsieh and his crew went on a cross-country tour that ended at that same worm farm, in Sonoma, California (see the "TerraVerso" profile in chapter 10 on page 211). Selling worms was the springboard for Tony's entrepreneurial success.

The founder of The North Face was wildly enthusiastic about vermicomposting. With 2½ million acres (1 million ha) in South America, Douglas Tompkins was the largest private landowner in the world. He owned dozens of organic farms and insisted that each have worm bins to turn agricultural waste into black gold. I visited Doug in Argentina and Chile in 2011.

You may be wondering how I got to know Doug Tompkins. Like thousands of other people throughout the world, he found me on the internet. When people search online for worms and vermicomposting, my name pops up. In July 2009 I received an email from someone in Argentina who wanted to create a state-of-the-art large-scale vermicomposting operation to produce humus for market gardens and orchards on a 7,500-acre (3,035 ha) farm. The message was signed Douglas Tompkins.

Although I had worked for a North Face subsidiary decades ago, I didn't recognize the name. People from over 100 countries have contacted me by email, so it was not surprising to hear from someone in Argentina. I assumed that Doug was a farmworker. Doug sent me several more emails during the next couple of weeks and then met with me in North Carolina in August. Imagine my shock when he revealed that he was founder of The North Face

FIGURE 0.1. At Doug Tompkins's Laguna Blanca organic farm in Argentina, the climate is mild enough for outdoor worm bins set up under this beautifully designed and built shelter from the sun.

and Esprit. I thought I would be chatting with a farmworker, and he turned out to be a multimillionaire!

A couple of years later, I accepted Doug's open-ended invitation to visit his organic farms in Argentina and Chile. Every farm used absolutely no chemicals and had very high yields. And each farm had a vermicomposting system to convert agricultural residues, animal manures, and food residuals to the finished product, called vermicast, which was then mixed with soil to boost the health of the soils and plants.

Tony Hsieh and Douglas Tompkins didn't make their *entire* fortunes by raising earthworms, although vermicast did boost Doug's income by increasing his crop yields. But I wanted to intensify your interest in worm farming by providing these examples of highly successful people who have enthusiastically promoted the value of earthworms and vermicomposting.

There are many reasons people decide to raise earthworms: to produce bait for their own use or to sell to others, to manage livestock manure or crop residues on farms, to process food scraps generated onsite at businesses and institutions, to produce vermicast to grow healthier food, to make an income from vermicomposting. All of these goals are feasible, but the first step is to learn how to raise worms and then buy a pound of worms and try it. You will develop earthworm husbandry skills and see how much you enjoy doing it. Maybe you won't like it and will decide to compost instead or do something else. But this way you will know if you have a knack for raising worms, with very little investment. My hunch is that you will love it and will want to expand what you are doing. This is the way to do it—start small and gradually grow your operation as you hone your earthworm husbandry skills.

I tell people that I became world-famous for vermicomposting expertise accidentally; it wasn't my intention. Although "Worm Woman" Mary Appelhof lived in my hometown, and we became friends while working together as community activists to establish recycling in our city during the 1980s, I resisted her invitations to become involved with vermicomposting. I was an undergraduate in college and Mary was 20 years older than I was when we met. My passion was working to reduce disposal of waste in landfills and incinerators via source reduction, reuse, and recycling. I wasn't interested in composting and vermicomposting back then.

In 1993 I started working at North Carolina State University as a waste management extension specialist. I was hired to help people understand how to recycle, because recycling had just been mandated statewide. My boss emphasized "publish or perish," so I developed proposals for seven extension factsheets. One of my manuscripts was titled *Worms Can Recycle Your Garbage*. My engineering colleagues sneered and made fun of me, saying, "You want to write about *worms*?" I had the last laugh, though, because the publication saw high demand and had to be reprinted several times.

That factsheet changed the course of my career. Calls and emails came flooding in from people wanting to know more about raising earthworms. I researched and created more publications to help satisfy their thirst for knowledge. I never intended to offer an annual training on worm farming, but people worldwide were constantly asking me, "When is the next conference?"

The NC State Vermiculture Conference is the only annual training event in the world that focuses on mid- to large-scale worm farming. It has attracted attendees from 30 countries. Participants learn how to set up or enlarge a vermicast production facility during this two-day event. Worm farmers describe their operations and discuss earthworm husbandry, feedstocks, equipment, monitoring, harvesting, and sales. Detailed marketing and business strategies are disclosed. Scientists share research about how vermicompost affects plant growth and suppresses plant pests and diseases. Over 100 worm enthusiasts from all around the world network, and many form business collaborations. Past attendees have come from 47 US states plus Argentina, Australia, Belgium, Canada, China, Dominican Republic, England, Greece, Guatemala, Hong Kong, India, Israel, Latvia, Mexico, Micronesia, Nicaragua, Nigeria, Norway, Philippines, Puerto Rico, Singapore, South Africa, Spain, Switzerland, Thailand, Trinidad & Tobago, Turkey, Uganda, and Zimbabwe. The conference website is composting.ces.ncsu.edu/vermiculture-conference.

For over 25 years most of my work at NC State University has been devoted to answering questions, writing publications, and presenting information about vermicomposting. In 2000 I established a 2-acre (0.8 ha) Compost Learning Lab (CL2) at NC State University's 1,500-acre (605 ha) Lake Wheeler Road Field Laboratory. The CL2 has a 40-by-30-foot (12 × 9 m) Worm Barn, an equipment shed, and a covered teaching shelter. There are 26 types of composting and vermicomposting bins and areas for hands-on training activities. People from all over the world visit CL2 to learn about vermicomposting and composting.

As the cofounder and a longtime member of the board of the North Carolina Composting Council (NCCC), I've also worked hard to help promote composting. See the resources section at the end of this book for more information about NCCC.

My goal in writing this book is to provide unbiased, experience-proven information about how to produce vermicast for profit or to manage waste produced by farms, community gardens, municipalities, businesses, institutions, and industries. You will learn the art and science of earthworm husbandry, and how to recycle waste to produce wealth. I start from the very basics, describing the benefits of vermicast and how it affects soil and plants, as well as key things to know about earthworm physiology and environmental needs. If you want to raise worms successfully, it's essential to know which species are right for vermicomposting and what they need to thrive.

After laying the groundwork, I turn to the practical matters of how to set up a vermicomposting business. It makes me sad that too many people rush into worm farming with very little knowledge and no realistic plan. They read some online articles and watch some videos and decide they know enough to start a worm farm. Often this information is inaccurate or misleading, and by following it you may fail to be successful. I offer my best advice about how to avoid common causes of worm business failure, and I cover regulatory issues as well as what to include in a business plan and a separate marketing plan.

The rest of the book gets into the specifics of setting up and operating a worm farm: choosing a vermicomposting system, what and how to feed earthworms, monitoring an active system and troubleshooting problems, harvesting vermicast, and options for

FIGURE 0.2. At the Compost Learning Lab, the shelter on the left overlooks 14 types of backyard composting bins. The white Worm Barn houses 12 worm bins ranging from household-sized to a 40-square-foot (3.7 m²) continuous flow-through bin.

selling or using it. I include some information about prices of vermicomposting equipment and vermicast products because that's something I'm frequently asked about when I give presentations. Keep in mind, though, that prices change over time and also vary from one part of the country—or the world—to another.

Throughout the book, you'll find plenty of practical examples from real-world worm farms and vermicomposting operations at schools and institutions. In the final chapter you will enjoy reading in-depth profiles of more than 25 vermicomposting operations in various parts of the world. I describe how people got started, how they expanded over time, and the variety of products they sell. Each has a unique vermi-operation, but they all possess an entrepreneurial spirit and desire to produce quality products that will positively contribute to the health of plants and people.

CHAPTER 1

Why Venture into Vermicomposting?

People from all walks of life are raising earthworms. Some begin vermicomposting as a hobby and scale up, whereas others have the intention right from the start to create a profit-making business. They may be farmers, community garden volunteers, entrepreneurs, landscapers, greenhouse growers, or staff at establishments that generate food waste. Some are retirees, and others work full-time. As you read the worm farm profiles throughout the book, you'll see that some large commercial vermicomposting operations were established by airline pilots, physicians, and dentists.

One of the most common reasons for venturing into commercial vermicomposting is the desire to recycle organic waste into products that enhance soil and plant health. Soil health has recently become an important international concern. The United Nations Food and Agriculture Organization declared 2015 the International Year of Soils to increase awareness of soil's role in food security and healthy ecosystem function. One objective of the campaign was to "educate the public about the crucial role soil plays in food security, climate change adaptation and mitigation, essential ecosystem services, poverty alleviation and sustainable development." At the end of 2015, the Paris climate agreement included a commitment from countries to increase soil carbon by 0.4 percent per year to help diminish global warming.

For decades soil was simply regarded as something that plants sprout from, and farmers used fertilizers augmented with herbicides and pesticides without regard for their environmental impact. Topsoil has now degraded so that much of it is unable to retain nutrients and water, and it is releasing carbon into the atmosphere (which then becomes carbon dioxide, a contributor to climate change). Chemicals in fertilizers are contaminating drinking water, and dust blown from degraded soil is causing respiratory ailments in people living in rural areas.

Vermicomposting is one action people can take to help do something about these alarming soil issues. The end product of vermicomposting improves soil health and fertility, increases the nutrient content and microbial life of soils, improves water retention, and reduces the need for fertilizers and pesticides.

Organic waste is material that originated from living organisms. Households, businesses, farms, industries, municipalities, and institutions all generate some form of organic waste, including food scraps, garden and lawn clippings, live-stock manures, paper products, brewery waste, pulp waste from food processors or paper mills, sewage sludge, textile mill fibers, or wood.

Organic materials such as food, paper, paperboard, and yard trimmings are the largest component of municipal solid waste generated in the United States. In 2015 they made up almost 55 percent of the waste produced. Food was the largest category of waste landfilled that year; it was a whopping 22 percent.

When organic waste is disposed of in landfills, it causes harm to health and the environment. As this waste breaks down in landfills, it releases liquid called leachate, which is toxic and can pollute groundwater, soil, and waterways. In the interior of a landfill, organic materials break down in an anaerobic environ-ment and release methane, a greenhouse gas that is 25 times more potent than carbon dioxide. The third largest source of human-made methane emissions in the United States is landfills.

A much better alternative to landfilling food waste and other organic materials is to vermicompost them. What was once considered waste can be transformed into valuable products that nourish soils and plants. Worm farmers have the power to solve environmental problems and turn a profit from their endeavors, too.

Vermicomposting, Vermicast, and Compost

Vermicomposting is a process that relies on earthworms and microorganisms to break down organic matter and transform its biological, physical, and chemical characteristics into a stable product that can be used as a valuable soil amend-ment and source of plant nutrients.

Vermicomposting turns organic materials into *vermicast*, which is a nutrient-rich, microbially active soil amendment or growth media for plants. When vermicast is added to soil, it boosts the nutrients available to plants and enhances soil structure and drainage. Vermicast helps plants grow bigger and produce higher yields, and it can reduce the impact of some pests and diseases.

For the past 25 years in dozens of documents, book chapters, and interviews, I have referred to the finished product of vermicomposting as *vermicompost*. This

FIGURE 1.1. Large-scale vermicomposting operations like this one in New Zealand generate tons of vermicast each year. *Photograph courtesy of Max Morley.*

term is often used to identify the mixture of earthworm castings (feces) and uneaten bedding and feedstock (organic material) that is harvested from worm beds. While some people use this term, others call it castings and still others call it vermicast. After considerable thought, I decided to use the term *vermicast* throughout this book, for two important reasons. One is because many people use the terms *compost* and *vermicompost* interchangeably, not realizing that the end product of vermicomposting is qualitatively different from compost. I think it would benefit the vermicomposting industry to distance itself from the term *compost* in referring to its products.

Vermicomposting and composting are very different processes, and it is important not to use the terms interchangeably. Composting is the controlled process of converting organic materials into a valuable soil amendment under aerobic conditions using biologically generated heat. In contrast, a vermicomposting pile or worm bin should be maintained so that it *does not* heat up. Too much heat can kill worms! In a compost pile the types and quantities of species of microorganisms change when the pile reaches thermophilic temperatures of 106°F (41°C) or higher. Temperatures in a worm bin remain in the psychrophilic or mesophilic range (well below 105°F/40°C), and thus there is a greater diversity and higher numbers of microorganisms during the vermicomposting process. The bottom line delineating the difference between compost and vermicast, though, is that the latter has passed through earthworms. Thus, vermicomposting is more similar to livestock production than to composting; it requires animal husbandry skills to properly care for the worms.

A second reason to use the term *vermicast* is to avoid product labeling that can be confusing to consumers. Some growers label their product castings and others, vermicompost. This inconsistency in labeling has also caused a bit of a rift among worm farmers. I like the way *vermicast* combines *vermicompost* and *castings* and leaves *compost* out of the term altogether.

Vermiculture Versus Vermicomposting

The terms *vermiculture* and *vermicomposting* are often used interchangeably, but they have different meanings. *Vermiculture* is the method of raising or breeding earthworms to sell for bait or food for fish, chickens, lizards, and other animals and reptiles. Growers focus on ideal conditions for worm growth, reproduction, and health. These worm farmers typically purchase and haul feedstock (food for worms) or pay for feedstock to be delivered to them, and they chiefly make money from sales of earthworms. An increasing number of them have realized that worm fecal waste is valuable, too, and they now sell vermicast in addition to earthworms. Moreover, some worm growers sell related products, such as worm bins, vermicast harvesters, shipping boxes, soil mixes, books, bedding, feed, and videos.

Vermicomposting is the process of turning organic debris into worm castings. The emphasis is on processing the waste rather than creating ideal conditions for raising earthworms. The size and reproductive rates of earthworms in a vermicomposting operation are frequently lower than those of the same species raised in vermiculture systems. Large vermicomposting facilities typically make money primarily from sales of vermicast. Some of them also make money by being paid to take waste that was destined for a landfill (vermicomposters typically charge lower rates than landfill tipping fees). Most of these operations do not sell earthworms, because their priority is to maintain an earthworm population to process the waste materials.

I am often asked, "How much castings do worms produce?" People assume there is an easy answer, as if science, mathematics, or engineering can cook up a formula: "If I feed X amount of food to my worms, how much castings will result?" The correct answer is: "There are too many variables to make a blanket statement about the output." Output depends on several factors, including the type of feedstock you are feeding the worms. (I go into detail about feedstocks in chapter 7.) If the feed has a high moisture content, then much of the water weight will evaporate or be absorbed into the earthworms' tissues during the vermicomposting process. Some additional variables are the type of system you are using, the temperature and moisture of the bedding, and the species of earthworm you are working with.

I've developed my own definitions of small-, mid-, and large-scale vermicomposting to use in my teachings and writings:

SMALL-SCALE VERMICOMPOSTING uses small bins and is practiced in homes and classrooms for the purpose of recycling food residuals.

MID- OR MEDIUM-SCALE VERMICOMPOSTING makes use of multiple or larger worm bins to process greater quantities of organic materials for use onsite or for sale. This is done at small farms and school gardens, onsite at businesses or institutions, or at homes in backyards, basements, or garages.

LARGE-SCALE EARTHWORM OPERATIONS have some of the same characteristics as mid-scale vermicomposting, but these operations create and follow a business plan, require higher monetary investment, and produce greater quantities of products for sale.

There are several good books available about small-scale vermicomposting; and thus this book focuses primarily on mid- to large-scale operations. Both mid-scale and large-scale operations utilize a variety of feedstocks, including food scraps, coffee grounds, brewery wastes, spent mushroom substrate, livestock manures, agricultural residues, yard debris, and other organic materials.

A Bit of Vermicomposting History

In the United States in the late 1800s, people began to realize they could make money by harvesting earthworms from farm fields and selling them as bait. In 1901 Shurebite Bait Company was established as the first business to sell worms.

In the 1930s Earl B. Shields wrote a booklet called *Making Money at Home*, which included a chapter called "Raising Earthworms." Requests for more information about this topic motivated Earl to learn more about it, and in 1951 he established a publishing business for books on earthworms. In 1951 Shields published *Earthworm Buyers Guide*, which provided advertising space for worm farmers to market their products. The directory continued to be published every two years until 2005, and the largest edition included 100 worm farms. In 1959 he published a book titled *Raising Earthworms for Profit: A Multi-Million Dollar Market* that has been reprinted numerous times. His publishing business sold more than a million of its 22 book titles. No doubt these books helped to stimulate growth in the worm farming industry.

In 1936 Dr. Thomas Barrett founded Earthmaster Farms in El Monte, California. A physician and researcher, Barrett investigated different methods for growing worms and published his findings in 1947 in his book *Harnessing the Earthworm*.

In the 1940s and '50s, several well-known people started worm farms. Perhaps the most prominent was Hugh Alton Carter Sr., a first cousin of former US president Jimmy Carter. Hugh succeeded Jimmy in the Georgia State Senate and held the seat for 14 years until he stepped down in 1981. After the Second World War, he started Hugh Carter's Worm Farm, growing crickets and worms for fish bait. They shipped bait countrywide and advertised as the largest worm farm in the world. Hugh Carter closed the business in 1996.

In 1967 a rabbit rancher named Ronald Gaddie bought worms to process the rabbits' manure. Before long he was making more money on worms than rabbits, so he established North American Bait Farm. His business continued to grow rapidly, and he established Bookworm Publishing Company to publish his own books and others. Along with Donald E. Douglas, Gaddie coauthored *Earthworms for Ecology & Profit*, volumes 1 (1975) and 2 (1977).

By 1972 Gaddie was grossing $100,000 in annual worm sales, and his income increased six times (to more than $600,000) over the next three years. The huge increase in sales was largely due to the buyback business model Gaddie adopted in 1972. He sold start-up packages to people who wanted to get into worm farming and pledged to buy back worms for a set price. However, the price per pound for the start-up package far exceeded the buyback price. And the worms Gaddie bought for the low buyback price he then sold to new customers for the high start-up price. Thousands of people bought into this "deal," and it inspired other less scrupulous people to establish similar pyramid schemes. Thousands of people were scammed until, in the latter half of the 1970s, the Securities and Exchange Commission shut down these buyback businesses, and Ronald Gaddie had to cease operations in 1980.

Worm farms sprang up all around the United States as a result of the pyramid schemes in the '70s. Some people lost their life savings on their investment when buyback contracts were no longer honored and they had no markets for their worms. However, many managed to survive after their sole market disappeared and thrived as they discovered the benefits of worm castings and began marketing them to farmers and other growers.

The worm farming industry has continued to grow over the past two decades, not just in the United States, but all over the world. The increasing popularity of organic farming and gardening has fueled interest in producing vermicast.

Who Is Vermicomposting Today?

Raising earthworms does not require exceptional skills or training, so it attracts a variety of people who want to use the worms for fishing or for profit. Thousands of people make part or all their income from earthworms. In

Vermicomposting's International Appeal

Vermicomposting is growing in popularity in many countries, and it is commonly practiced in Cuba. Why is it so popular? After the collapse of the Soviet Union in 1991, which resulted in the termination of Soviet-Cuban relations, Cuba lost 80 percent of its imports overnight, and the Cuban people were forced to become self-reliant in producing food. With the loss of fertilizer imports, worm farms quickly popped up to produce organic fertilizer. By 2016 there were 172 worm farms in Cuba. The most commonly used vermicomposting feedstocks are cow manure and filter cake, a waste produced when processing sugar.

Anna de la Vega visited Cuba in 2016 to learn about their organic agriculture and tour worm farms. Anna lives in England and is the director of a business called The Urban Worm Community Interest Company. She earned a Winston Churchill Memorial Trust fellowship that enabled her to visit worm farms in the US and Cuba. Anna gave a fascinating presentation about her travels at the 2017 NC State Vermiculture Conference.

One of the organic farms Anna visited in Cuba was called Organopónico Vivero Alamar. It is a 26-acre (11 ha) urban farm located in Havana. The farm had six worm bins constructed of concrete blocks, each 49¼ feet (15 m) long by 9¾ feet (3 m) wide by 23½ inches (60 cm) deep. The worms were fed horse and bull manure every 15 days in a layer 4 inches (10 cm) deep. The farmers harvested the bins every 60 days, producing 300 tons (272,155 kg) of vermicast per year.

FIGURE 1.2. These worm bins are at Organopónico Vivero Alamar in Cuba. *Photograph courtesy of Anna de la Vega.*

developing countries where income opportunities are limited for women, they are achieving financial independence from earthworms and sales of vermicast.

During the past eight decades, many people have raised earthworms as a source of income or as a means of managing organic waste. Some are drawn to the business by extravagant claims of vast potential markets for earthworms in large waste disposal systems and agriculture, and as a source of food for animals. Despite these claims, the current major commercial use of earthworms is as bait for freshwater sport fishing. An increasing market for earthworms is selling them to others who wish to vermicompost, ranging from sales of a pound or two for beginner backyard "worm farms" to hundreds of pounds to large-scale commercial operations.

A growing number of people are getting into vermicomposting because they want to produce vermicast for their own use rather than buying it. Producing their own gives them control over the source of all of the feedstocks and materials and assurance that the vermicast is a high-quality product.

Journalists and folks who want to get into the vermi-business often ask me how many commercial worm farmers there are in the United States. They are always disappointed when I tell them that no one knows. The vermi-industry is not organized and has no trade group. Some large producers in the United States have private contracts and no need to advertise for more business, so even their neighbors don't realize there's a large worm farm nearby. Some worm farmers keep a low profile because they don't want to be discovered by state regulatory officials. And some producers try not to attract attention because they think their vermi-system is uniquely effective and don't want others to steal their ideas.

What I can tell you is that vermicomposting is increasingly being adopted at businesses, institutions, farms, and municipalities for managing organic waste. They are interested in implementing sustainability practices while at the same time cutting costs. Organic materials are being vermicomposted primarily onsite, though in some instances they are transported to a centralized facility. Some businesses and institutions have purchased commercially manufactured worm bins, and others have designed and built their own. Some operations have bins set up directly on the ground; others stack bins on pallet racks. Vermicomposting can help save the cost of fees for solid waste collection and disposal, water usage costs, and wastewater disposal expenses.

Vermicomposting at Businesses

Businesses that vermicompost food waste include restaurants, grocery stores, hotels, food processors, wholesale food outlets and farmers markets, shopping malls, resorts, retirement communities, paper manufacturers, and office buildings with dining facilities. In 2014 businesses generated 36 percent of the total food wasted in the United States, according to the Environmental Protection Agency (EPA).

An indoor vermicomposting spot at a business can be set up in a basement or break room. Kona Pub & Brewery in Hawaii was focused on sustainability and promoting a "green image"—they placed worm bins right in their dining room! Worm bins can be located outdoors on loading docks, alleys, or shady areas on lawns.

The Bittercreek Alehouse & Red Feather Lounge bar/restaurant in downtown Boise, Idaho, began their vermicomposting project over a decade ago. They dove right in, purchasing two specialized vermicomposting bins called continuous flow-through reactors, which they set up in the restaurant basement. The restaurant was able to vermicompost their food wastes, coffee grounds, and paper menus and beer lists. They worked up to a capacity of about 300 pounds (135 kg) of material per week. When the vermicast was ready, the restaurant staff harvested it and put it through a screening device called a trommel (to remove large particles that weren't fully digested). They sold or traded the vermicast to local farmers who supplied produce to the restaurant. To learn more about the Bittercreek Alehouse vermicomposting operation, see the profile in chapter 10.

Even an athletic stadium can do a little worm farming on the side. In 1995 I visited the Seattle Kingdome athletic stadium because I heard it had a vermicomposting setup. Sure enough, there were 12 worm bins that vermicomposted 50 pounds (23 kg) of food preparation salad waste per week. Four of the bins were handmade of wood, and the other eight were constructed from 55-gallon (210 L) barrels cut in half lengthwise and laid on their sides. Leaves and shredded paper were used for bedding and to cover the food waste. Vermicast was harvested every few months and used on the Kingdome's flower beds.

Vermicomposting at Institutions

The types of institutions that generate food waste include hospitals, schools, universities, prisons, military bases, daycare centers, nursing homes, long-term-care facilities, residential and summer camps, and government centers. The EPA estimated that 8 percent of food waste generated in the United States in 2014 originated from institutions such as educational facilities, prisons, and hospitals.

One example of a long-standing vermicomposting project at a college is the organic farm at the Evergreen State College in Olympia, Washington. Beginning in the early 1990s, they vermicomposted food waste from dining services and a local restaurant in hand-built wooden bins. In 2008 they switched to a modular continuous flow-through bin system. The vermicast is used on the college farm and community gardens.

Vermicomposting at Schools

Although schools are institutions, they are worth discussing separately because for several decades many classrooms throughout the United States have used worm bins as tools for lessons in science, mathematics, and writing, and even art and drama. Several people (including myself) have created curricula for classroom worm bins to be taught by teachers and cooperative extension agents.

Numerous schools expanded from classroom-scale worm bins to school-wide vermicomposting of lunchroom food scraps. This required approval and cooperation among administration, teachers, cafeteria staff, custodians, students, and parents. Often these projects were initiated by teachers or parents, but sometimes students were the catalysts, asking, "Why is food being thrown away? Can't we recycle it, like we do with paper, cans, and bottles?"

A rapidly growing trend is for schools, from preschools through colleges, to maintain gardens onsite. Students learn how plants grow and harvest vegetables and fruits that are then consumed in the lunchroom. It's a natural connection to compost or vermicompost the food residuals and use the products to nourish the plants to grow more food.

FIGURE 1.3. This open-air shed at Salmon Creek Elementary School in the Freestone Valley of California is framed with lumber milled from trees grown onsite. Shade cloth on the sides keeps conditions cooler for the vermicomposting bins inside.

California has the greatest number of schools in the United States that are vermicomposting their cafeteria food residuals. That is not surprising, because they have a mild climate and can easily maintain outdoor worm bins. Sonoma County is particularly active with school-wide vermicomposting. This is largely due to the efforts of the Compost Club, which was founded in 2003 by a concerned parent named Rick Kaye (who also set up vermicomposting at The Bishop's Ranch). With a group of volunteers, the Compost Club, now known as the Worm Wizards of Waste, established school-wide vermicomposting at 27 schools in Sonoma, Napa, and Marin Counties. In chapter 10 I include the full story of the Compost Club and its amazing projects, which by May 2015 had diverted a total of 169 tons (153,315 kg) of food waste from landfills (see page 181).

Vermicomposting at Farms

There's a growing trend for farms such as dairy farms and hog farms to vermi-compost manure and crop residuals as part of the farm operation. Farmers are choosing to vermicompost for several reasons. Some need an environmentally beneficial alternative for manure management. Others want to produce vermi-cast to increase their crop yields and reduce their use of fertilizers, herbicides, and pesticides. Still others want to put old, unused animal housing to productive use. And some farmers choose vermicomposting to increase their income from the sales of earthworms or vermicast.

Vermicomposting by Municipalities

It is uncommon in the United States, but some municipalities are vermicomposting food residuals, yard debris, or sewage sludge. Municipalities may operate vermicompost facilities on their own or contract with a private entity.

From 2004 to 2009 the city of Middleton, Connecticut, vermicomposted a total of 4½ tons (4,080 kg) of food scraps from their city hall, a university, library, restaurant, and juvenile detention center. Using grant funding, the city constructed a greenhouse and bought eight worm bins, each 8 feet long by 4 feet wide by 1 foot high (2.4 × 1.2 × 0.3 m). The worm bins were stacked on pallet racking to maximize vertical space. They collected pre-consumer food waste in 5-gallon (19 L) buckets with lids. They mixed the food with an equal amount of shredded leaves in wheelbarrows. The leaves helped to absorb excess moisture, prevent odors, and create a better balance of nutrients for feeding worms. The mixture was applied daily to the worm beds in shallow layers and covered with a 2-inch (5 cm) layer of either compost, leaves, or paper.

One example of a public-private partnership is the arrangement between an Australian firm called Vermitech and two townships in Pennsylvania, Granville

Big Red Worms

Jeremiah Picard owns this vermicomposting business in Lincoln, Nebraska, which is also a form of public-private partnership. Jeremiah's business operates a Community Composting Initiative: Residents pay a small yearly fee to drop off their food waste at any of 15 locations. He also picks up food waste from seven schools and several local businesses, including coffee shops and restaurants.

For food waste collection, Jeremiah provides schools and businesses 64- or 48-gallon (240 or 180 L) Toter rolling carts. Coffee shops receive a 30-gallon (115 L) rolling bin because coffee grounds are so heavy.

Individual members package their food waste in compostable bags that are tied off and placed in the pickup bed, which can hold 2,500 pounds (1,135 kg). Food waste collection is on Monday and Wednesday for some locations, Tuesday and Thursday for others. Every school and business in the group has its food waste collected by Big Red Worms on Saturday mornings.

During the past three years, Jeremiah has diverted more than 1 million pounds (453,500 kg) of food waste and almost 2 million pounds (907,000 kg) of horse manure and bedding from the landfill.

FIGURE 1.4. Jeremiah Picard of Big Red Worms collects food waste around the community in this pickup truck, which has an insert with a cart tipper installed in the bed. *Photograph courtesy of Kate Picard.*

and West Hanover, to vermicompost their sewage sludge. The sewage and water authorities turned to Vermitech to compost their biosolids after the local landfill closed. West Hanover spent $3 million to construct a 40,000-square-foot (3,715 m²) area housing a 21,000-square-foot (1,950 m²) building with geothermal climate control that could hold 14 continuous flow-through bins. The first two worm beds began operating at the site in 2008. Dewatered sludge cake with 10 to 20 percent solids was mixed with wood chips and fed to worms one or two days per week. Finished vermicast was harvested once a week and retrieved with a specialized conveyor belt system.

Vermicomposting by Entrepreneurs

Starting a vermicomposting business can take many forms. Many folks start small and then scale up fairly quickly, but keep a tight rein on start-up costs by using low-cost materials, avoiding mechanization, and developing do-it-yourself skills. One example is the Texas Worm Ranch in Garland, Texas. Heather Rinaldi is the sole owner of the "ranch," which is actually located at an industrial park. Although most of her business activities take place inside a 12,000-square-foot (1,115 m²) warehouse, she uses space outdoors for preparing feedstock, pre-composting, and test gardens. Even though Heather has a large-scale vermi-operation, she has no capital loans; she was able to avoid debt by making her bins and screening systems by hand.

In the mid-2000s Heather began researching soil microbiology, because she was interested in creating healthy soil to make wholesome food. She started a worm bin in 2008, and when she harvested the vermicast, she made tea from it and used it on her plot at a community garden. Within two weeks the plants in her plot looked far better than ones in other plots, so her fellow gardeners started asking Heather for vermicast tea and worms.

Heather increased her vermicomposting activities to meet the demand and rented a warehouse in 2010 to accommodate her operation. By 2012 she needed a larger space, so she moved into a 6,000-square-foot (555 m²) warehouse. The next year she doubled the size of her business and moved into the 12,000-square-foot warehouse she currently occupies. She has one full-time staff member and three to five part-time workers.

Heather's business model has two primary objectives. The first is continuous quality improvement, measured by biology. Producing a density and diversity of biology in her vermicast is her goal. "My goal is not to produce the most vermicompost that I can; it's to produce the best," says Heather. Her second objective is education. She wants to teach people how to improve their health by eating nutrient-dense food that contains fewer chemicals.

Her method of vermicomposting is based on observing nature. She uses three types of feedstocks for bedding and feed for her worms. Wood trimmings are dropped off by tree trimmers, leaf waste is delivered by lawn-care companies, and food waste is picked up at juice bars and vegan restaurants. Area landscapers and tree services drop off their waste in piles outside Heather's warehouse. About 2 tons (1,815 kg) of yard waste, leaves, and wood chips are delivered each week.

Rather than pre-composting, Heather lets the piles mature for several months before introducing the materials to her indoor worm bins. She doesn't blow air into the piles or turn them. Instead she spices up the piles with vermi-cast left over from tea brewing, coffee grounds, and pumpkins that are dropped off after Halloween. The piles of organic material are inhabited by earthworms and microorganisms that help decompose the matter.

Inside the warehouse Heather has 90 worm bins that are 8 feet long by 4 feet wide (2.4 × 1.2 m). The bins are made of wood that was recovered from a local building project. The bedding used to start the bins is composted material from the outdoor piles. The feedstock is pre-consumer fruits and vegetables from a nearby juice bar. The feedstock is covered with leaves to hold in moisture and prevent flies. Vermicomposting inside the beds takes 8 to 12 weeks; however,

FIGURE 1.5. The microscope lab at Texas Worm Ranch where they check the microbiology of each worm bin. *Photograph courtesy of Heather Rinaldi.*

FIGURE 1.6. Worm bins inside the warehouse at Texas Worm Ranch. *Photograph courtesy of Heather Rinaldi.*

FIGURE 1.7. Texas Worm Ranch worms. *Photograph courtesy of Heather Rinaldi.*

if the bins are set up with processed material from other bins as bedding, the process goes faster.

When it is time to harvest, Heather's staff empty a bin by shoveling it out by hand, dumping the material on a handmade mobile sifting machine that they pull up to the worm bin. The material lands on a ¼-inch (6 mm) landscape screen hanging from a rectangular frame built of 2×4s. A motor (which cost $450) shakes the frame to cause the finer material to fall through. The staff calls this a pre-sifting procedure to separate the bigger chunks of wood from the worm-processed matter.

The finer vermicast is then rolled in a trommel screener that—you guessed it—they built themselves. The trommel, dubbed Bertha, is 8 feet long and 30 inches wide (2.4 × 0.8 m). Bertha is suspended inside a huge "bathtub" (made of tarps) that catches vermicast that passes through the ⅛-inch (3 mm) screen and channels it into two containers underneath. A variable-speed motor turns the trommel.

Heather's husband, Vince, is an engineer. He learned all about cycle time management, efficient and replicable processes, and quality control maintenance in college. He helped Heather put these concepts into practice to produce products with consistent quality. For example, her worm bins are on a rotational cycle time. A percentage of bins is set to be harvested each week, according to the length of time needed for vermicomposting, so that they can harvest enough fresh vermicompost for weekly demand by customers. Every week there are 5 to 10 bins ready for harvest. Heather plans her vermicast sales based on when the materials will be ready, so they are fresh. Within two weeks of harvest, her vermicast is with her customers.

Her customers include Major League Baseball training fields, grass-fed beef producers, hay farms, organic tree-care companies, and gardeners. During spring and fall customers pick up 800 to 2,000 pounds (365–907 kg) of vermicast per week. Many of them make their own tea from Heather's vermicast to spray on lawns, trees, and hay fields. The customers bring back the 25-gallon (95 L) plastic tubs they used the week before to refill them.

She ships vermicast only to people who desire her high-quality product and are willing to pay the shipping fees. "I'm not in the business to ship mass quantities of castings at a low price," says Heather.

For backyard gardeners and households, she bags castings in 30-pound (14 kg), 10-pound (4.5 kg), and 2-pound (0.9 kg) bags. She uses plastic, breathable sand bags that are refillable.

She has a homemade 80-gallon (305 L) brewer with a 1,700-gallon-per-minute (6,435 L) pump that actively aerates tea. She used to brew a lot of tea for customers, but she is trying to cut back on it and encourage people to brew their own. She will brew to order; the customer must pick up the tea.

Heather's goal for 2018 is to produce 20,000 pounds (9,070 kg) of vermicast per month. During January through May, she sells 100 to 200 pounds (45–91 kg) of worms per month. The rest of the year, she sells 50 to 100 pounds (23–45 kg) monthly. Heather's focus is not on worm sales, and she doesn't advertise the fact that she sells worms.

Heather loves to teach and has a classroom at her facility. One or two Saturdays a month, she holds classes on vermicomposting, gardening, or soil microbiology. Her classes include hands-on activities in the outdoor compost piles and inside the worm warehouse. For $50, an adult participant gets a two-hour class and goes home with a 10-gallon (38 L) Rubbermaid tote with ½ pound (0.2 kg) of worms. Children participate in hour-and-a-half classes and for $20, they get a shoebox-sized bin with worms and seeds to plant in a 4-inch (10 cm) pot.

The city of Middleton, Big Red Worms, and the Texas Worm Ranch are just a tiny snapshot of what's happening in the world of mid- and large-scale vermicomposting operations around the United States. Throughout this book you'll find more real-life examples of vermicomposting facilities, and chapter 10 provides in-depth descriptions of more vermicomposting setups at businesses, public institutions, and public schools, as well as profiles of commercial worm farming operations.

Markets for Earthworms

The largest market for earthworms by far is bait anglers, as about 50 million people go fishing in the United States each year. There are plenty of smaller markets, too. People buy worms to set up bins at home to process their kitchen food scraps. Farmers and commercial vermicomposters make large purchases of earthworms for their operations. Game bird breeders, tropical fish stores, frog farmers, fish hatcheries, pet stores, zoos (with exotic fish and birds), and poultry growers buy earthworms to feed to their animals. People who educate the community, such as teachers, extension agents, and recycling coordinators, often need to buy earthworms for setting up new worm bins. Universities, private laboratories, and high schools buy earthworms for research and class-room needs. Worms are being used to remediate soils contaminated with heavy metals. Earthworms are also being used as active ingredients in pharmaceuticals, cosmetics, and skin-care products.

If you live near or along a route to fresh- or saltwater fishing spots, you can sell earthworms directly to fishing enthusiasts (if local zoning laws allow this commercial activity). You can also ship them by mail or other delivery services (see chapter 9, page 159); place advertisements in national magazines directed to fishing enthusiasts and home gardeners; or advertise on the internet.

Worm farmers can sometimes sell earthworms to locally owned sporting goods or fishing tackle stores, but most of the larger stores of this type purchase bait from established wholesalers. Additional establishments that sell bait include marinas, hardware stores, gas stations, convenience stores, and general goods retailers such as Walmart.

Increased interest in vermicomposting has created a high demand for earthworms, and there are reports of earthworm growers not having enough stock to fill orders. These worm-hungry markets include:

- New commercial vermicomposting facilities
- Worm farms that are expanding their operations or replacing dead worms
- Livestock farms, rabbitries, and horse stables
- Businesses and institutions that want to vermicompost cafeteria waste or other organic materials
- Industries that wish to turn their organic wastes into value-added products by vermicomposting them (including food processors, paper mills, breweries, cardboard manufacturers, cotton mills, land reclamation sites, sludge/biosolids producers, canneries, and wineries)

Markets for Vermicast

Many worm growers focus on selling vermicast rather than earthworms. The biggest market for vermicast to emerge in recent years is sales to marijuana growers. The majority of states have legalized medical marijuana, and a growing number have legislated approval for recreational purposes. Thus, marijuana markets have increased exponentially. And it appears that most pot growers value the beneficial effects of vermicast on plants and soil, so demand for it has expanded. An increasing quantity of scientific research that demonstrates the beneficial uses of vermicast is assisting the growth of market outlets. Vermicast is increasingly being used by organic gardeners and farmers. Other market outlets are garden supply stores, home improvement centers, landscape contractors, nurseries, greenhouses, grocery chains, flower shops, discount houses, golf courses, vineyards, resorts, and the general public. Vermicast is being sold in bulk by the cubic yard or in bags. Sometimes it is mixed with compost and soil blends.

Some worm growers also market vermicast tea or liquid extract. They extract organic matter, microorganisms, and nutrients from vermicast to produce the tea. The liquid extract can be applied as a soil drench or used directly on plant foliage. Worm growers sell vermicast tea at farmers markets or to orchards, nurseries, greenhouses, farms, landscapers, and retailers. I discuss this in detail in chapter 9 on page 170.

Earthworm-Related Commerce

In addition to selling earthworms and vermicast, many worm farmers sell a diverse assortment of products and services to supplement their incomes. You can also promote your worm farm by applying your logo to T-shirts, hats, water bottles, coffee mugs, tote bags, and the like.

Supplies and educational materials. If your target customers are new earthworm growers, you can purchase supplies at wholesale prices and market your business as one-stop shopping. Supplies you could stock include bait cups, shipping cartons, bedding, peat moss, and feeds. Additional items you can sell are how-to guides, newsletters, books, videos, worm bins, and harvesting equipment.

Baits and fishing equipment. If you sell earthworms for bait in a retail business, you can also offer other types of live bait such as mealworms, grubs, nightcrawlers, baby frogs, crickets, grasshoppers, or crawfish. You can raise

FIGURE 1.8. The Worm Farm in Durham, California, has developed a wide range of vermicomposting-related products including bins, accessories, and books that they sell at their facility and through their website. *Photographs courtesy of Mark Purser.*

these baits yourself or buy them wholesale. You could also sell tackle and fishing equipment such as hooks, lures, lines, sinkers, and floats.

SOIL AMENDMENTS. In addition to vermicast, you can sell compost and other soil amendments such as coconut coir, peat moss, potting soils, bat guano, greensand, and perlite.

EARTHWORM COCOONS. Earthworm cocoons (sometimes called encapsulated earthworms) are easier to ship than live earthworms. About three worms will hatch from each cocoon in 32 to 73 days, depending on temperature, humidity, and moisture.

VERMICAST TEA. Sales of vermicast tea have significantly increased over the past decade. You can sell these liquid extracts or vermicast and tea brewing systems. For more information about vermicast tea, see chapter 9, page 170.

WORKSHOPS AND CONSULTING SERVICES. Offering workshops for new worm growers can be a great source of income, with participation fees ranging from $25 to $1,000. You can offer workshops for children in classrooms, summer camps, after-school programs, and youth group meetings. Homeowners, teachers, and college students also like to attend vermicomposting workshops. Many worm growers offer consulting services to help others get started in the business.

CHAPTER 2

How Vermicast Benefits Soils and Plants

T housands of scientific studies on vermicompost have been conducted worldwide. I just went online to Google Scholar and typed in the word *vermicompost*. In 0.03 second the website displayed 27,500 articles and studies that have been published in scientific peer-reviewed journals. Scientists have extensively documented the effects of vermicast on plant growth, health, and yield and identified the specific mechanisms behind these effects. They have also discovered that vermicast helps plants resist attacks by pathogens, insects, and parasitic nematodes. If you are interested in reading some of these studies, check out the bibliography at the end of this book, where I have listed many fascinating and useful studies about the effects of vermicast on plant growth, and disease and pest suppression.

What Is Vermicast?

First let's return to the question, "What is vermicast?" In chapter 1, I said that vermicast is a mixture of earthworm castings and uneaten bedding and feedstock harvested from worm beds. This means that we started with organic materials for bedding and added feedstocks. The worms consumed most of the food and bedding and left behind a mixture of their castings (worm poop) and undigested organic materials.

Let me clarify that although *vermicast* is used as a general term, the vermicast from my worms is not identical to the vermicast from your worms, and your vermicast is going to be different from that of other producers. That's because the nutrient content of vermicast (and its electrical conductivity, bulk density, moisture, pH, hormone and humic acid content, et cetera) depends on the types of feedstocks and bedding you provide for your worms in addition to how you raise them.

I feed cow manure to the worms in my vermicomposting systems, and if you also feed your worms cow manure, you might assume that your vermicast is the same as mine. However, the manure I use comes from dairy heifers that plop it straight into wood shavings. The manure-and-wood-shaving mixture is then shoveled into a truck and delivered to my site, where it sits in the open in a field for a few months. I feed the mixture to my worms once a week, until the pile is depleted and my worms need more.

If you say, "I feed my worms cow manure," first, I will wonder if it is dairy or cattle manure. These two types of cows are raised for distinct purposes, so their feed and care are quite different, and the manure they produce will thus have differing consistencies. Similarly, grass-fed cows raised in pastures will produce different manure from cows raised inside a building. In large operations, cow manure is often washed out of buildings with copious amounts of water. If you want to use that manure, you will have to buy or build a machine to separate

FIGURE 2.1. Finished vermicast like this produced by Earthen Organics is a dark, nutrient-rich soil amendment. *Photograph courtesy of Kristen Beigay.*

the solid manure from the liquid. If manure lands in bedding, however, it will be mixed in with that organic material, which will further differentiate it as a feedstock. The bedding may be sand, wood shavings, sawdust, paper, compost, hay, wood chips, straw, peanut hulls, corn stubble, cardboard, tobacco stems, or some other industry by-product.

Unlike me, you might not feed raw cow manure to your worms at all. Perhaps you mix rice hulls or silage or some other carbonaceous material with your manure and then pre-compost it (pre-composting is discussed in chapter 7 on page 124).

So you see? When cow manure is the feedstock, the resulting vermicast will differ depending on what the cows are raised for, how they are raised, what the cows are fed, whether the manure is mixed with water, what type of bedding or carbonaceous material it may be mixed with, and whether it is raw or pre-composted.

How do different types of vermicasts affect plants? Differently, of course! Some degree of generalization is necessary, however, so let's compare cow manure vermicast to vermicast made from hog manure, paper waste, and food waste. The Soil Ecology Laboratory at The Ohio State University used these four types of vermicast for numerous studies, comparing their effects on plant growth and yields, germination rates, and pest and disease repellency. As you might guess, the composition of nutrients, hormones, and humic acids in these four different types of vermicast had different effects on specific plant species. For example, some plants reacted more positively to hog manure vermicast than to paper waste vermicast, and it was the opposite for other types of plants. The differences aren't necessarily drastic, but they're significant enough that if you or your customers are raising a specific crop, such as strawberries or tomatoes, you may want to feed your worms the feedstock that will produce the vermicast with the most favorable effects on that particular plant species.

Is your head spinning? That's how I feel when I try to explain these concepts during a presentation on vermicomposting. There are many variables to keep track of and evaluate. The take-home message is simple, though: Different feedstocks and methods of raising worms will produce distinct vermicasts, and these vermicasts will have a range of effects on plants.

Vermicast Versus Compost

The price difference between conventional compost and vermicast is astronomical. Whereas thermophilic compost usually sells for about $30 per cubic yard (0.8 m³), the same amount of vermicast may be sold for prices ranging from $200 to $1,800. Yep, if you sell vermicast instead of thermophilic compost, you will get paid 7 to 60 times more for it.

The Persuasive Power of Packaging

You may be wondering why a customer would pay $1,800 per cubic yard for vermicast instead of $200. The secret is the packaging. If you put a small amount of vermicompost into a tea tin or some other type of attractive packaging and sell it for $4.99, $9.99, $14.99, or even $19.99, people will buy it because it seems inexpensive. What they do not realize is that if they purchased a cubic yard's worth of that cutely packaged vermicompost, it could cost $1,800. This is one example of the importance of smart marketing, which I cover in chapter 4 on page 60.

There is a good reason for the price difference between vermicast and thermophilic compost. As I mentioned in chapter 1, the types of microbial communities that develop during the composting and vermicomposting processes are significantly different. While thermophilic microorganisms predominate in conventional composting, mesophilic microbes—those that thrive in moderate temperatures—prevail during vermicomposting. And, the microbial populations in vermicast are substantially larger and more diverse than those in thermophilic compost.

Vermicast contains higher nutrient concentrations than conventional compost, but the soluble salt levels are lower. Scientific studies show that vermicast contains lesser amounts of soluble salts and greater cation exchange capacity than the parent material consumed by the earthworms. This is an advantage for plants because greater cation exchange capacity in soil generally makes it easier for plant roots to absorb nutrients.

A hot environment, such as a thermophilic compost pile, releases more nitrogen into the atmosphere than a cooler process, like vermicomposting. Consequently, vermicast typically retains higher nitrogen levels than conventional compost, and because nitrogen is the nutrient most likely to be deficient in soils, this provides vermicast with a distinct advantage over compost.

Several studies have also shown that vermicast has significantly lower carbon-to-nitrogen (C:N) ratios than compost. One example demonstrating these differences is summarized in table 2.1, which presents the results of a study conducted at the University of Hawaii that analyzed 157 samples of vermicast and compost from various sources over a period of five years. Among other results, take note that the C:N ratios were significantly lower in the vermicast samples. These lower C:N ratios make nutrients more available to plants.

I should also note that vermicomposting can take as little as half the time of the thermophilic composting process. After active decomposition, thermophilic

TABLE 2.1. Select Nutrients Found in Vermicompost and Thermal Compost

Method	Feedstocks	Number of Samples		Nutrient				PPM	
				N (%)	C:N	P (%)	K (%)	NO₃- N (%)	NH₄- N (%)
Vermicompost	Food waste	42	Mean	2.11	12:1	0.79	1.47	1672.2	141
			Range	0.89– 4.59	5:1– 25:1	0.06– 2.06	0.06– 4.83	267– 2986	2– 969
			CV (%)	39	27	70	98	93	164
	Manure	59	Mean	1.67	14:1	3.04	0.55	1988.88	185
			Range	1.29– 2.25	10:1– 18:1	0.40– 6.03	0.08– 2.39	316– 4824	1– 2063
			CV (%)	15	15	53	111	72	255
Thermal Compost	Green wastes	28	Mean	1.28	21:1	0.45	0.77	634	21
			Range	0.67– 2.72	8:1– 40:1	0.14– 0.92	0.21– 1.12	35– 1913	0– 175
			CV (%)	37	42	47	30	80	369
	Manure and mortalities	28	Mean	1.77	19:1	1.69	1.68	3834	243
			Range	0.61– 3.01	10:1– 26:1	0.21– 3.78	0.48– 3.13	60– 8625	26– 1813
			CV (%)	42	47	70	58	79	163

Note: **CV:** Coefficient of variation is an indicator of the variation within the compost samples for that nutrient.
N = nitrogen; **C:N** = carbon/nitrogen ratio; **P** = phosphorus; **K** = potassium; **NO₃-N** = nitrate nitrogen;
NH₄-N = ammonium nitrogen
Source: Radovich and Arancon, *Tea Time in the Tropics*

compost needs to be cured for a couple of months or longer, whereas vermicast does not need to be cured and can be used immediately after it is harvested.

Characteristics of Vermicast

As explained in chapter 1, vermicast is stabilized organic matter produced under controlled conditions by the interaction of earthworms and microorganisms. But people sometimes get confused about the difference between organic matter and organic material. *Organic material* originates from plants and animals. *Organic matter*, on the other hand, is organic material that has been decomposed by microorganisms, a process that changes its structure and reduces its mass by up to 90 percent. Thus, feedstocks for worms are organic materials, and the worms (plus microorganisms) turn those organic materials into organic matter in the form of vermicast.

Organic matter causes mineral soil particles to bunch together into clumps, forming soil aggregates that protect against erosion and improve overall structure. This increases permeability (water infiltration) and helps soil to absorb and hold moisture, thus making it more resilient against drought and heavy rainfall. Organic matter also reduces soil compaction and surface crusting and improves the tilth of soil.

In addition to possessing all of the benefits of organic matter described above, vermicast has a near-neutral pH and contains plant growth hormones, humic acids, fulvic acids, and an extraordinarily large and diverse microbial population.

Vermicast is granulated and has a fine particulate structure. It has a high surface area and water-holding capacity, and contains nutrients in plant-available forms, including nitrates, soluble potassium, transferable phosphorus, exchangeable calcium, and magnesium. Vermicast releases these nutrients gradually in a balanced release pattern.

The humic and fulvic acids in vermicast dissolve insoluble minerals and make them available to plants. These acids stimulate plant growth, help plants establish healthier roots, and increase yield. Vermicast also contains plant hormones and plant-growth-regulating substances, including auxins, gibberellins, cytokinins, ethylene, and abscisic acid.

Vermicast Effects on Soils

New approaches to agriculture have the potential to improve crop production, human health, and the environment. Land-grant universities are working with growers and the US Department of Agriculture to test innovative techniques for better managing soil. Three new primary practices that promote soil health have emerged—all of them directly opposed to the decades-old farming practices that relied on use of synthetic fertilizers, herbicides, and pesticides. Instead of disturbing the soil with tillage, no-till practices are being adopted. In place of planting the same crop over and over, rotational cropping is being encouraged to increase soil fertility and reduce soil erosion. Instead of leaving soil bare between rows or after harvest, farmers are planting legumes, vegetables, or grasses to cover the soil, thus returning nutrients to the soil, reducing erosion, and supporting biodiversity.

Vermicomposting, too, can play an important role in the remediation of soil. Research studies have shown that vermicast improves soil aeration, porosity, and water retention. The reason vermicast improves the physical structure of soil has to do with its enhanced microbial populations and activity, absorbent organic matter, polysaccharides, and mucus secretions that help cement soil particles together, causing aggregate stability.

Bountiful Benefits for Soil

Incorporating compost or vermicast into soil provides the following benefits:

- Adds organic matter
- Helps soil to absorb and retain water
- Breaks up clay soils
- Improves soil structure
- Increases cation exchange capacity
- Eases cultivation
- Helps form soil aggregates
- Enhances soil fertility
- Reduces bulk density
- Improves soil aeration
- Increases soil microbial populations
- Reduces soil compaction
- Diminishes soil erosion
- Reduces pH
- Lowers electrical conductivity
- Helps prevent soil crusting
- Provides micro- and macronutrients and increases their availability

Vermicast Effects on Plant Growth

Studies show that amending soil with vermicast causes seeds to germinate more quickly, enhances rate of seedling growth, increases root numbers and biomass, improves root stress tolerance, leads to earlier flowering of plants, and increases plant yields. Vermicast decreases plant transplant shock and increases plant vitality and flavor profile. Plants grown with vermicast have more leaves and flowers, more total leaf area, greater plant biomass, and higher leaf chlorophyll content. What's more, all this usually occurs independent of nutrient availability.

What do I mean by "independent of nutrient availability"? Here's an example. About 15 years ago a couple of my colleagues and I received a grant to study the nutrient runoff of vermicast-amended soils. We set up randomized plots in three different fields during fall and spring seasons for a year and a half, including control plots with no vermicast added, plots with the addition of 10 percent by volume vermicast, and plots with the addition of 20 percent by volume vermicast. Each plot received the same amount of nitrogen fertilizer (as determined by the amounts already existing in the soil, the levels of nitrogen in the two vermicast applications, and the turnip plants' needs). We installed devices to collect rainfall runoff and analyzed the nutrient content to determine how the addition of vermicast affected runoff water quality. Although we were not studying the effect of vermicast on plant growth, we couldn't help noticing the astounding differences between turnips with no vermicast amendments, and those with 10 percent and 20 percent by volume.

FIGURE 2.2. These three turnips received equal amounts of nitrogen. The turnip, *left,* received no vermicast, while the turnips, *center* and *right,* were grown in soil amended with 10 percent and 20 percent (by volume) vermicast. *Photograph courtesy of J. Mark Rice.*

We took a quick photo of three representative turnips, which is what you see in figure 2.2. Look at how much more extensive the roots are on the two larger turnips, in addition to the size differences. This is a dramatic example of how vegetable growers can hit the jackpot simply by adding 10 percent by volume vermicast to the soil.

Note that all three turnips received equal amounts of nitrogen. Nitrogen is the most important nutrient needed by plants, so if different amounts of nitrogen are not influencing the plant growth, what is? Although vermicast contains a variety of crucial macro- and micronutrients, it appears that the following elements in vermicast have an even greater effect on plant growth: humic acids and plant growth hormones such as gibberellins, cytokinins, and auxins.

My friend Dr. Norman Arancon, a professor of horticulture at the University of Hawaii, has reported that seedlings germinate and plants flower two weeks earlier if they are grown with vermicast. He says this is due to hormones, not nutrients. Scientific literature contains numerous studies of vermicast that demonstrate increased germination, growth, and yield in a variety of vegetables, crops, and ornamental plants in greenhouse, lab, or field conditions. The plants studied include tomatoes, bell peppers, okra, potatoes, cucumbers, strawberries, cherries, peas, raspberries, spinach, corn, mung beans, marigolds, petunias, coriander, eggplants, sorghum, sunflower, hyacinth beans, poinsettias, bachelor's button, impatiens, gerbera daisies, grapes, chrysanthemums, lettuce, wheat, sugarcane, rice, pumpkins, and cowpeas.

How much vermicast is being used in these experiments? Scientists have reported plant growth increase by incorporating 5 to 50 percent by volume vermicast in soil mixes. At application rates of 60 percent or higher by volume, plant growth is affected negatively. Dr. Arancon states that vermicast provides the greatest benefit to plant growth when it makes up 10 to 40 percent of the growing medium by volume.

Although many effects of vermicast are independent of nutrient availability, vermicast also does make two important nutrients more available to plants: nitrate and calcium. Nitrate is the type of nitrogen most needed for healthy plant growth. Many studies have shown that vermicomposting promotes nitrification—the conversion of ammonium-nitrogen to nitrate-nitrogen.[1] Earthworms have calciferous glands that excrete calcium carbonate into worm castings. Calcium is vital to plants for building strong cell walls; it also enhances the absorption of nitrogen.

Arthropod Pest Control

When vermicast is added to soil, it can make plants resistant to arthropod insects. Vermicast is rich in phenolic compounds, which are repellent or toxic to many types of insects, because the phenolics taste bad and can disrupt insect digestive systems. If plant roots can uptake and metabolize the phenolics contained in vermicast, the plants will acquire the ability to significantly deter arthropod insect attacks.

The scientific literature contains several greenhouse and field studies demonstrating vermicast's ability to suppress a variety of insects. Norman Arancon, Clive Edwards, and their colleagues have shown that vermicast can deter the following pest attacks on plants:

- White caterpillars and aphids on cabbages
- Two-spotted spider mites on eggplants and bush beans
- Mealy bugs, two-spotted spider mites, and aphids on tomatoes and cucumbers

Further examples of arthropod pest control in the scientific literature include E. N. Yardim et al.,[2] who used vermicast in field and greenhouse experiments to reduce damage to cucumbers from spotted and striped cucumber beetles and to repel tobacco hornworms from tomatoes, as well as K. R. Rao,[3] who found that groundnuts raised in vermicast-amended soil experienced fewer attacks by aphids, spider mites, jassids, and coccinellid beetles.

The list doesn't end there, either. Vermicast has been proven to effectively suppress attacks by other insects, including larval hornworms, thrips, European corn borers, fruit borers, brinjal fruit and shoot borers, corn pests, leaf miners, and psyllids.

Plant Pathogen (Disease) Control

Dr. Arancon has conducted a tremendous number of scientific studies on how vermicast affects plant growth and suppresses diseases, arthropod pests, and parasitic nematodes. Norm has been a speaker at my annual vermiculture conference about 15 times. Norman was a postdoc with Dr. Clive Edwards at The Ohio State University and conducted many of the studies in their Soil Ecology Laboratory. Since 2008 he has been teaching and continuing his research as a professor at the University of Hawaii–Hilo.

Norman has conducted research on vermicast suppression of the following plant diseases: *Pythium* on cucumbers, *Rhizoctonia* on radishes, *Verticillium* on strawberries, *Phomopsis* and *Sphaerotheca fuliginea* on grapes, *Phytophthora capsici* on tomatoes, *Fusarium oxysporum* on peppers, *Plasmodiophora brassicae*, *Phytophthora nicotianae*, *Fusarium lycopersici*, and *Rhizoctonia solani*.

When Dr. Arancon gives a presentation at my vermicomposting conference, he shows dozens of bar graphs that depict the effects of vermicast and teas on plants. One of my favorites is "*Pythium* Suppression by Vermicomposts," which is shown in figure 2.3.

The graph shows pythium suppression in cucumber seedlings that were planted in a soil-less medium called Metro-Mix 360 (MM360), which is commonly used in experiments and consists of sphagnum moss, peat, composted bark, and vermiculite. For this particular experiment they sterilized the MM360, added all of the plants' needed nutrients, and inoculated all of the seedlings with the pythium disease. The control contained no vermicast, while other treatments had 10 percent, 20 percent, or 40 percent vermicast added by volume. The experiment had an additional twist: Half of the vermicast treatments used sterilized vermicast, and the rest were unsterilized. The chart shows a dramatic difference between the sterilized and unsterilized treatments! The first bar shows the control with a Disease Severity Rating (DSR) of close to 3.5 (on a scale of 1.0 to 4.0). The second set of bars depicts sterilized vermicast treatments of 10 percent, 20 percent, and 40 percent by volume, and indicates that the sterilized vermicast did little to suppress the pythium disease. The last group of bars shows the effects of 10 percent, 20 percent, and 40 percent

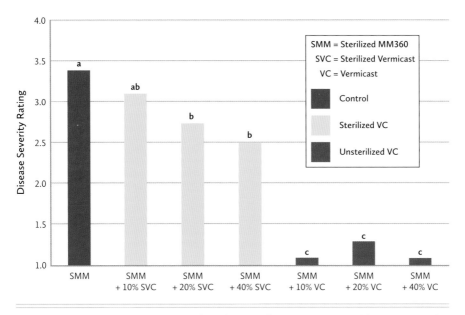

FIGURE 2.3. This graph shows how the addition of vermicast to a soil-less mix provides excellent suppression of the fungus that causes pythium rot in cucumbers. *Source:* Arancon, "*Pythium* Suppression by Vermicompost," Ohio State University.

treatments with unsterilized vermicast. I have three words to describe it: "Bye, bye, pythium!" It's another knock-your-socks-off image, like my turnip study photo, where you can clearly see the dramatic effect that vermicast has on plant disease suppression. The 10 percent and 40 percent unsterilized treatments produced the same DSR ratings: 1.1 (on the 1.0 to 4.0 scale), more than 2 points lower than the control. This graph clearly shows that the abundance of beneficial microorganisms in unsterilized vermicast treatments effectively suppresses the pythium disease.

At the Research Institute of Vegetable Crops in Poland, Dr. M. Szczech conducted a similar experiment with sterilized and unsterilized vermicast treatments for the infection of *Fusarium oxysporum* f. sp. *lycopersici* on tomato plants. The vermicast lost its effectiveness after heating (sterilization), and the author concluded that the large numbers of antagonistic bacteria and fungi in the unsterilized vermicast were suppressive to the pathogen.[4]

I've just described research demonstrating the effectiveness of unsterilized vermicast in suppressing disease, and the relative ineffectiveness of sterilized vermicast. This discrepancy occurs because the sterilization process kills off most of the microorganisms in vermicast. As I've already discussed, the vermicomposting process maintains and supports a large diversity of mesophilic microorganisms, which grow best in moderate temperatures, typically between 68 and 113°F (20–45°C). This range is significant because some countries have already adopted (or are considering) regulations to sanitize finished vermicast. They recommend sterilizing the vermicast at a temperature of 158°F (70°C) for one hour in order to destroy pathogens. That temperature will kill viruses and parasites if the mass of vermicast is completely penetrated, but it will also eradicate most of the microbial population, thus leaving a narrow array of microbial species in significantly reduced numbers. That will drastically reduce the effectiveness of vermicast.

I should note that there are a handful of cases where vermicast does not function as an effective pathogen control. Dr. Arancon has tested the effects of vermicast produced from different feedstocks (such as dairy manure, paper waste, and food residuals) on a wide variety of plants after infecting them with diseases, and although his research has quite consistently shown that vermicast suppresses disease, there are some exceptions. For example, when he inoculated ginger with the disease *Ralstonia solanacearum*, some of the treatments produced symptoms of disease on the plant foliage or in the soil. However, even with the presence of the disease in the soil, the vermicast increased yield in the ginger plants.[5]

To determine whether vermicast would be an effective disease suppressant for a particular crop, you can reference the numerous plant disease suppression studies in the bibliography at the end of this book or go to Google Scholar and type into the search bar, "vermicompost plant disease suppression."

I will end this section with anecdotal evidence from a large vermicast manufacturer in California. He reported that local vineyards were losing 25 percent of their vines due to disease. After adding vermicast, these vineyards are now only losing 1 to 2 percent of their vines!

Plant Parasitic Nematode Control

Many studies have shown that vermicast application suppresses plant parasitic nematodes such as the root-knot nematode, *Meloidogyne hapla*. Some scientists believe that adding vermicast to soil increases populations of nematodes' microbial enemies, including *Trichoderma* spp., *Pseudomonas* spp., chitinolytic bacteria, and *Pasteuria penetrans*. It is also thought that vermicast causes an increase in the number of arthropods that eat plant parasitic nematodes, such as Collembola and nematophagous mites.

Additional research has shown that vermicast amendments promote the growth of fungi that can annihilate cyst nematodes and enlarge populations of rhizobacteria, which generate enzymes toxic to plant parasitic nematodes.

Dr. Norman Arancon and his colleagues have conducted various studies on the effects of solid and aqueous vermicast on plant parasitic nematode attacks.

They found that root galls formed by *Meloidogyne hapla* on tomato plants were suppressed by the addition of food waste vermicast treatments of 10 percent, 20 percent, and 40 percent (by volume) in MM360. In that particular experiment the 20 percent vermicast treatment had the greatest effect.[6] They have also demonstrated plant parasitic nematode suppression on grapes (*Vitis vinifera*) and on strawberry field crops (*Fragaria × ananassa*) with additions of vermicast made from paper and from food waste.[7]

CHAPTER 3

Key Things to Know About Earthworms

I don't want to put you to sleep with a long-winded discussion of the anatomy and physiology of earthworms. Instead, in this chapter I focus on the fundamentals you need to know to perform successful animal husbandry skills that will help your earthworms flourish. If you find yourself becoming fascinated by earthworm physiology and behavior (some people do!), there are lots of great books and scientific articles where you can pick up plenty of fascinating earthworm trivia.

Anatomy and Physiology Basics

Earthworms are terrestrial invertebrate animals with a similar physical structure in all species. They belong to the phylum Annelida (meaning "ringed" or "segmented") and the class Oligochaeta (small bristled worm). All of the vital organs are contained in the front third of an earthworm's body. The rings around the body delineate segments, and each segment has muscles and bristles, called setae, to aid in movement. The number of segments varies among earthworm species—for example, nightcrawlers have about 150 segments, and red wigglers have around 95. The setae anchor a segment of the worm to whatever surface it is on, while muscles and setae in other segments move the earthworm in whichever direction it wants to go. The segments lengthen or contract independently to move the worm.

Earthworm bodies are like tubes with no protruding sense organs or appendages, and this enables them to move easily through soil. They do not have eyes or ears, but their brain connects to nerves attached to their skin and muscles.

Their nerves sense light and vibrations, and their muscles move in response to the stimuli. Earthworms have well-developed circulatory, digestive, nervous, excretory, muscular, and reproductive systems. The anterior end (head) has a tonguelike prostomium lobe, and the posterior end has an anus.

Earthworms have simple circulatory systems. Instead of a heart, they have five pairs of aortic arches that pump blood through blood vessels. Dorsal blood vessels transport blood to the front of the earthworm's body, and ventral blood vessels carry blood to the back of the body.

The following are the basic characteristics of earthworms that worm farmers need to know. Earthworms are:

- Cold-blooded
- Hermaphroditic
- Without lungs, so they breathe through their skin
- Susceptible to dying if their skin dries out
- Light sensitive; exposure can cause paralysis within about one hour

What does it mean to be cold-blooded? Basically, you can't regulate your body temperature, so your body assumes the same temperature as the environment. If it is very cold, you cannot shiver to raise your body temperature. If it is very hot, you cannot sweat to cool your body temperature. You can die if the temperature is too high or low. As temperatures rise higher or drop lower than their comfort zone, earthworms' activity will diminish and they will consume less food.

A hermaphrodite is an animal that has both male and female reproductive organs. Does that mean that it can self-reproduce? Although occasionally one earthworm will produce babies without mating, there are no records of the babies surviving. It takes two earthworms to tango—that is, to produce viable offspring.

The gaseous exchange of oxygen and carbon dioxide occurs through the skin of earthworms. Their skin needs to stay moist to allow the passage of dissolved oxygen into their bloodstream. Earthworms can release body fluid and mucus to coat their skin and keep it moist, but they still need a damp environment to survive and thrive.

Although earthworms do not have eyes, they have receptor cells that can detect different intensities and wavelengths of light. Earthworms do not respond to red light, so if you want to study their activities without disturbing them, you should observe them in the dark under red light. They will move away from blue light (and other colors of the spectrum) if they can, and it is especially important to shield earthworms from bright light, which can induce a state of paralysis. If earthworms are exposed to light for too long (about an hour), they will become paralyzed and die when their skin dries out.

Amazing Earthworm Facts

Believe it or not, there are more than 9,000 species of earthworms, and they range in size from ½ inch to 22 feet long (13 mm–6.7 m). You are probably thinking that "22 feet long" is a typo, since it's hard to imagine a worm that is even 22 inches long. But it's not. The longest earthworm on record was discovered in South Africa in 1967 stretched across both lanes of a two-lane road. It was an unusually large member of the species *Microchaetus rappi*, which typically grow to 6 feet (1.8 m) long and ¾ inch (19 mm) in diameter. There are other cases of supersized earthworms, such as:

THE GIANT GIPPSLAND WORM (*Megascolides australis*) in Australia can grow to 13 feet (4 m) long, and averages 10 feet (3 m) in length and ¾ inch thick.

THE PRUSSIAN BLUE (*Terriswalkeris terraereginae*) is found in the mountains near Cairns, Australia. It grows to 6 feet long, looks like a giant gummy worm, and leaves behind a trail of glowing mucus.

THE NORTH AUCKLAND GIANT (*Spenceriella gigantea*) grows to 4½ feet (1.4 m) long and is bioluminescent (it glows in the dark). It is said that you can read by their light.

THE WASHINGTON GIANT EARTHWORM (*Driloleirus americanus*) is also called the giant Palouse earthworm and grows to 3 feet (91 cm) long and ½ inch (13 mm) thick. These albino worms can burrow 16 feet (4.9 m) deep, give off a lily scent when handled, and spit in self-defense.

THE OREGON GIANT EARTHWORM (*Driloleirus macelfreshi*) grows to 4 feet (1.2 m) long.

Ecological Groups and Species

While you are wrapping your mind around giant earthworms and the existence of thousands more worm species than you ever imagined, get ready for this: *Some earthworms don't live underground.* Scientists divide earthworms into three groups according to where they live and what they eat. The first group is called anecic (*an-EE-sik*). They are the earthworms we are most familiar with—big, fat nightcrawlers. They live in underground vertical burrows and eat soil and leaf litter. The next group is endogeic (*en-doe-JEE-ik*), which is made up of smaller earthworms that live in horizontal burrows and consume large quantities of soil. Both of these groups do what we expect *earth*worms to do, right? They live in the soil and eat the soil. But there is a third group of earthworms called epigeic (*ep-i-JEE-ik*), which live in, and consume, litter. These worms subsist on decomposing organic material such as leaf litter and livestock manure, so you will likely find them in leaves, manure, or compost piles. It is very important to remember that epigeic earthworms do not want to live in soil.

Only seven species of earthworms have been identified by scientists as suitable for vermicomposting. Four are temperate species, and three are tropical.

TEMPERATE SPECIES

- *Eisenia fetida* (red wiggler)
- *Eisenia andrei* (red tiger)
- *Eisenia hortensis/Dendrobaena veneta* (European nightcrawler)
- *Lumbricus rubellus* (red earthworm)

TROPICAL SPECIES

- *Perionyx excavatus* (blue worm)
- *Eudrilus eugeniae* (African nightcrawler)
- *Amynthas gracilis*, which is sometimes mistakenly called *Pheretima hawayana* (Alabama or Georgia jumper)

Some researchers have identified *Amynthas gracilis*, *Lumbricus rubellus*, and *Dendrobaena veneta* as invasive earthworms in the United States, and they do not recommend them for vermicomposting. These species have the ability to move into previously uninhabited areas and disturb forest ecosystems. Although *Eisenia fetida* and *E. andrei* are also nonnative species, they are not invading and harming forests and do not have the potential to do so because they die when exposed to freezing temperatures.

About *Eisenia fetida*

Eighty to 90 percent of commercial worm farms use *Eisenia fetida* (eye-SEN-ee-a FEH-tid-a) earthworms for vermicomposting. I use the scientific name because there are numerous common names for *E. fetida*, such as red wiggler, tiger worm, redworm, manure worm, California redworm, composting worm, brandling worm, panfish worm, and trout worm. People get confused and think these are different species of worms. To avoid any misunderstandings about what we are talking about, let's use the one scientific name that is assigned to this species of earthworm: *Eisenia fetida*. I should note, however, that there is a bit of confusion surrounding the spelling of the scientific name due to spelling changes made by earthworm taxonomists. You see, this earthworm species was originally called *Eisenia fetida*, but scientists later changed the spelling to *Eisenia foetida*. You will often see the latter spelling in literature, on the internet, and in scientific articles, but taxonomists actually changed the spelling back to *Eisenia*

FIGURE 3.1. *Eisenia fetida* earthworms are the worm of choice for vermicomposting, and they are a desirable species for use as fishing bait, too. *Photograph courtesy of Kristen Beigay.*

fetida several years ago, and this remains the correct name. So keep that in mind and spell it correctly, and be sure to write it in italics with the first word capitalized and the second word lowercase.

Eisenia fetida is the preferred earthworm for vermicomposting because they can withstand a wider range of environmental conditions, such as temperature and pH, than other worms. They are more efficient breeders than many other species of earthworms, and they are described by worm growers as the "most adaptable and hardy" of cultivated earthworm species. Many worm farmers who raise *E. fetida* for bait prefer them to nightcrawlers because they are less temperamental, reproduce faster, and can be shipped without refrigeration. They also claim that this species stays alive longer on a fishhook and produces an odor that lures fish.

For a sustainable vermicomposting system, you should start with at least 1 pound (0.5 kg) or about 1,000 *Eisenia fetida* per 1 square foot (0.09 m²) of surface area of the worm bin or pile. It's important to use surface area in your calculation, rather than overall volume, because worms are top feeders and they will mostly live in the top 4 to 6 inches (10–15 cm) of organic material in the bin.

Although it might seem like an easy way to cut costs, it is not wise to grab wild earthworms from your yard after a rainstorm, because you will have no idea what species they are. Even if you take worms from a manure or compost pile,

Invasive Earthworm Species

Invasion of nonnative earthworms! Sounds alarming, doesn't it? Occasionally people will see headlines like this and get worked up about invasive earthworms, and in recent years people have asked me to address the issue when I give vermicomposting presentations. I assure them that *Eisenia fetida* are not associated with any of the invasive earthworm problems you hear about in the news. Like many earthworm species, *E. fetida* arrived in the United States hundreds of years ago when Europeans first migrated here. They require warm, moist organic matter and cannot survive outside of worm bins, compost piles, or livestock manure piles. They don't burrow in soil and won't survive in a garden or lawn. All of the scientists who are studying earthworms' effects on forests agree that *E. fetida* are not to blame and are incapable of causing this harm. Dr. Lee Frelich, at the University of Minnesota, is a leading researcher on this topic. Dr. Frelich concurred that *E. fetida* (red wiggler) earthworms are *not* a problem because they won't invade the natural environment.[1]

they will likely be a mixture of species, and they may not get along in confinement. In addition, it will be difficult to know exactly how many you have. Do you really want to count out 1,000 or more worms from a manure pile?

"What about a bait shop?" you might ask—after all, that's the place most people think of when it comes to buying worms. However, bait cups usually contain only 28 to 40 *Eisenia fetida*, so you would need 25 to 36 bait cups to reach a total of 1,000 worms. In my local area, bait cups with an average of 28 worms cost $3.38 per cup, and on the internet I've seen 40 red wigglers in a bait cup selling for $8.99. Based on these prices, purchasing 1,000 worms in bait cups would cost you $122 to $225! Plus, what would you do with those empty 25 to 36 Styrofoam or plastic cups?

It is best to start your vermicomposting operation by purchasing *Eisenia fetida* earthworms from a worm grower. The cost will be $20 to $50 per pound, and that may or may not include shipping. If you buy significantly larger quantities of earthworms, you may be able to pay a lower price, say $15 per pound. But be careful! Some worm growers are honest, some are honesty-challenged, and some simply don't know what they're doing. Certain growers may not know what species of worms they have, and others may intentionally sell you a species that isn't *E. fetida*. I've also encountered growers who do not weigh their worms carefully and will send you far fewer than 1,000 when you pay for a pound. When I buy a pound of *Eisenia fetida*, I don't count every worm, but I visually check to see if it seems to be about 1,000 worms. Because such problems in the supply chain do exist, I advise you to inquire carefully to find worm growers that seem to have a good reputation. To start, purchase only 1 pound from a grower, and figure out whether they have given you the correct species and quantity. If everything checks out and you want to buy more worms for a larger system, you will feel confident buying more worms from that particular grower.

Earthworm Reproduction

The presence of a band (called a clitellum) around a worm's body indicates the ability to reproduce. The eggs and sperm of each earthworm (remember, they are hermaphroditic) are located in separate parts of the body to prevent self-fertilization. Two worms will mate by facing in opposite directions and connecting their clitella, which secrete mucus that encircles both worms. While the worms are locked together, they exchange sperm. Once they have completed the exchange, the worms disengage and their movements push their mucous bands toward their heads. As the band moves over the earthworm's body, it picks up sperm and mature eggs that were deposited by the other worm. The mucous tube will slip over the earthworm's head, and both ends will close into what is called a cocoon. Inside the cocoon, the sperm and eggs unite to create worm embryos that develop into baby worms. The cocoon is golden brown, shiny, and looks like a tiny lemon the size of a match head. Mature *E. fetida* can produce two or three cocoons per week. Two to seven (on average, three) *Eisenia fetida* babies will hatch from each cocoon, crawling out in about 30 to 75 days, depending on environmental conditions. Newborn worms resemble whitish pieces of thread less than ¼ inch (6 mm) long. Their skin progressively darkens, and within a few days they look like baby reddish-colored worms.

FIGURE 3.2. These two *Eisenia fetida* earthworms are locked together to mate.

Earthworms reproduce quickly when food and water are available. After the babies hatch, it takes 53 to 76 days for them to reach sexual maturity. In the right conditions, they can live for four to five years or even longer. When food or water are not accessible, cocoons will stay dormant and not hatch for months to years; this gives them the capacity to survive long-term even in unfavorable conditions.

Earthworm Needs for Vermicomposting

Eisenia fetida earthworms need moisture, oxygen, food, and moderate temperatures. Other variables for a healthy worm bin include pH, ammonia, salts, bulk density, earthworm population mass, and microbial density and diversity.

Moisture

Moisture content is the percentage of total weight of a material that is water. Because earthworms require adequate moisture to help them breathe through their skin, worm beds need to maintain a moisture content range of 60 to 85 percent. An ideal target is 80 percent moisture.

Depending on the size and type of your vermi-system, you may need to put some type of covering over the bed to sustain adequate moisture. See chapter 8 for more information about determining moisture levels and maintaining the proper moisture.

Oxygen

Earthworms can survive in relatively low-oxygen and high-carbon-dioxide environments, and they can even stay alive when submerged in water if it contains dissolved oxygen. However, if there is no oxygen, earthworms will die.

The amount of oxygen available to worms during vermicomposting can be affected by the amount of feed, moisture, and depth of their bedding. Low-oxygen conditions may occur if a worm bin is too wet (over 90 percent moisture), the bedding is too deep, or there are excessive amounts of decomposing feed in an enclosed space.

Food

The earthworm's digestive tract extends the whole length of its body. Earthworms have tiny mouths and no teeth, so they can't take a bite of

FIGURE 3.3. When conditions are right, earthworms will thrive—as they do in this bedding mix from a vermicomposting operation in New Zealand. *Photograph courtesy of Max Morley.*

food like we can. They can only manage to consume microorganisms and tiny bits of organic matter that can fit into their minuscule mouths. The earthworm's prostomium pushes food into its mouth, and the pharynx (throat) sucks it in. The food gets coated with saliva, and then muscular contractions push it down the esophagus; this contains calciferous glands that release calcium carbonate, which neutralizes acidic decaying food. Instead of going into a stomach, the food travels into the worm's crop, where it is stored and mixed, and then on to the gizzard, where it is crushed and ground apart by tiny stones or bits of grit that were swallowed by the worm. Next, food is moved into the intestine, where it is further broken down by digestive enzymes. The digestive fluids release amino acids, sugars, bacteria, fungi, protozoans, nematodes, and other microorganisms, in addition to partially decomposed plant and animal materials from the food the worms have swallowed. Simpler molecules of food are absorbed through intestinal membranes and absorbed into the bloodstream for cell production and energy use by the earthworm, and the rest passes out the anus as mucus-coated castings (worm poop). Castings are teeming with microorganisms as well as undigested plant material, nutrients, and soil. It takes two and a half hours for food to pass through a worm. The worm's digestive system is not very efficient—only 5 to 10 percent of the food taken in is absorbed.

Temperature

Eisenia fetida earthworms live comfortably and breed at temperatures between 55 and 85°F (13–29°C). They can tolerate temperatures between 40 and 90°F (4–32°C), but will have lower activity at 40 to 50°F (4–10°C) and try to move to a warmer area. Conversely, they will move to a cooler space to get away from excessive heat.

For commercial earthworm production, the ideal temperatures for growth and activity range from 60 to 80°F (16–27°C). Bed temperatures between 60 and 70°F (16–21°C) facilitate intensive cocoon production and hatching.

pH

The pH of organic materials indicates whether it is acidic (1 to 6), neutral (7), or alkaline (8 to 14). Earthworms will grow in a pH range between 5 and 8, though acidic conditions can create problems in the worm bed. For commercial production, you should keep your earthworm beds at a pH range of 6.8 to 7.2. You can measure pH by either using litmus paper (a colorimetric system), using a pH meter and electrode, or sending a sample to a laboratory for analysis.

Ammonia and Salts

Ammonia gas can form in a worm bed if the pH of the bedding is over 7.5 and the ammonium level is high. (Ammonium is found in animal urine and fresh manure.) This gas can kill earthworms, especially if their bins are too tightly covered to allow proper ventilation. Earthworms are also sensitive to salts, so it is important to be aware of the salt levels of feedstocks and to prevent salts from accumulating in the bed. This is one advantage of the type of vermicomposting system called continuous flow: Earthworms can continually move into fresh feedstocks, so there tend to be fewer problems with the accumulation of salts.

Bulk Density

Bulk density is the weight of organic material in a given volume. It increases with compaction and tends to increase with depth. Bulk density impacts oxygen transfer and moisture in the bedding and, in turn, the ability of the worms to move around. At higher bulk densities less air and moisture will be diffused through the bedding, and worms will be inhibited from moving about. Lower bulk density means greater porosity and air spaces. When bulk density is higher than 1,000 pounds per cubic yard (454 kg/0.8 m³), the organic material is heavy and pore space is decreased, making oxygen diffusion increasingly difficult. The ideal bulk density rate is between 800 and 1,000 pounds per cubic yard (363–454 kg/0.8 m³).

Population Density

Eisenia fetida earthworms can be found anywhere in a worm bin, but the majority will usually stay within the top 4 inches (10 cm) of bedding. They may move deeper into the bedding if conditions are less than adequate. The minimum population density for successful vermicomposting should be 1 pound of worms per square foot of surface area (0.5 kg/0.09 m²). However, a healthy, active vermi-system can contain 2 to 3 pounds of worms per square foot of surface area (0.9–1.4 kg/0.09 m²).

Microbial Density and Diversity

Microorganisms play a critical role in vermicomposting systems. To survive, these microbes need moisture, oxygen, food, and a hospitable temperature and pH range. Bacteria are by far the most abundant microorganisms in a vermicomposting system and are responsible for most of the decomposition taking place. Fungi are also found in high numbers in a worm bin, and they can decompose organics that are low in nitrogen or too dry or acidic to be taken care of by bacterial activity. Bacteria break down simple carbon compounds, but fungi can degrade tougher organic materials that contain cellulose, hemicellulose, and lignin. Other types of microbes found in worm bins include protozoans, rotifers, and actinomycetes.

Microorganisms eat the organic material in the system, and earthworms then consume the microbes. The microorganisms need carbon to fuel their metabolism and nitrogen for cell synthesis and to make enzymes that decay the organic materials. The way the microbes "eat" is by secreting enzymes that dissolve organic materials via breaking down chemical bonds. Enzyme activity increases as temperatures rise, but if it gets too hot, enzyme activity will decrease. With every 18°F (10°C) rise in temperature, from 32 to 95°F (0–35°C), there is a 1.5 to 3 percent increase in microbial activity.[2] Enzyme activity also decreases if temperatures drop too low.

Microorganisms need proper proportions of carbon and nitrogen for effective cell growth. Carbon supplies energy and growth, and nitrogen provides protein and supports reproduction. Microorganisms require about 25 times more carbon than nitrogen. The ideal ratio is 25 to 30 parts of carbon to 1 part nitrogen (25:1 to 30:1), but a range of 25:1 to 40:1 is reasonable. When the carbon-to-nitrogen ratio is 20:1 or lower, nitrogen can form ammonia gases, which harm the worms. At a carbon-to-nitrogen ratio higher than 40:1, the degradation process is slowed due to limited nitrogen for microbial cell growth.

Water is also necessary for microorganisms to stay alive. It supports their metabolic processes and provides them with a way to move around. When organic material has a moisture content less than 40 percent by weight, microbial activities slow down.

The Vermicomposting Process

Vermicomposting maintains a wide diversity of mesophilic organisms throughout the process. Earthworms ingest microorganisms; some are digested, while others pass through the gut unharmed. The grinding action of the earthworm's gizzard promotes microbial activity in the feed as it passes through the gut, and indigenous microflora in the gut of *Eisenia fetida* further contribute to the microbial community in their castings. Mucus from the intestinal tract covers castings as they are expelled, which provides a carbon source for soil microbes and leads to a flush of microbial activity in fresh castings. In chapters 6 and 8, I explain in detail how to manage the vermicomposting process. But before we get into the nitty-gritty of managing worm bins or beds, it's important to lay the groundwork by learning about government regulations that relate to vermicomposting and formulating a sound business and marketing plan.

Earthworm Companions

When you start vermicomposting, you'll quickly discover that earthworms aren't the only living critters in the worm bed. It's important to get to know these other invertebrates that feed on decomposing organic matter in worm bins and what role they play. They likely hitched a ride on the organic materials you added to the bin, such as manure, food residuals, soil, decaying leaves, or paper. They will not harm your worms, so don't worry about them or try to control their populations. These decomposers are part of the soil food web and will actually improve the productivity of your worm bed. I'll briefly describe a few of the nonharmful decomposers you may find in your worm bed.

WHITE WORMS (ENCHYTRAEIDAE)

These small, white, segmented animals, also called pot worms, are related to earthworms. Threadlike and less than an inch (2.5 cm) long, they are often mistaken for baby worms. Their numbers may be greater in moist, acidic conditions, but they won't harm your worm workers or outcompete them for food.

SPRINGTAILS (COLLEMBOLA)

These important soil organisms may be the most numerous perceptible decomposers in your worm bin. They have the ability to jump and are so small that they look like salt granules hopping about.

SOW BUGS AND PILL BUGS (ISOPODS)

Sow bugs and pill bugs are covered with flattened plates; they resemble miniature armadillos when viewed through a hand lens. Also called wood lice, or slaters in Australia, they are usually brown or gray and ½ inch (13 mm) long. They are no threat to your worms and help decompose organic matter.

MILLIPEDES (DIPLOPODA)

These arthropods have pairs of legs on most of their body segments. They will live harmoniously with earthworms and other decomposers in your worm bed.

CHAPTER 4

Planning for Success

This chapter is not a comprehensive tutorial on how to write a business plan, because that could be a book in itself. Instead it's a collection of my best advice about business planning, including helpful resources (even some low-cost or free ones), and my insights on the aspects of business planning that are especially relevant for worm-related businesses. And the chapter also offers tips for marketing worms and vermicast.

Business planning is a complex undertaking that requires hours and hours of research, thinking, and writing. And if your reaction as you read the chapter is, "I have no idea how to do some of this stuff!" (such as writing contingency plans), that's a sign that you need to gain more experience or knowledge before you launch a business. You may need to visit some worm farms to see how those businesses are run, or attend a workshop or conference, or work with a consultant.

Much of the information in this chapter won't apply if you want to produce vermicast only for use on your own farm, or to nonprofit institutions and schools. However, complying with regulations, the first section of the chapter, is important for *everyone* planning a mid- to large-scale vermicomposting operation to read, because some regulations apply even to those who do not sell any products.

The take-home message here is to develop your knowledge, experience, and skills before starting a vermi-business. Start small, and expand later if you desire. Write a detailed business plan, make conservative estimates, set realistic goals, and create backup plans. Create a comprehensive marketing plan based on selling diverse products to a variety of customers. Secure enough funding to cover all of your costs, plus enough to help you recover from initial missteps or unforeseen adversities. Plan for marketing and promotional costs, and put a lot of time and energy into building a strong customer base. Be willing to acquire new products or make modifications according to your customers' desires. And above all, *be honest*—with yourself, partners, lenders, regulators, suppliers, and clients.

Check State and Local Regulations

When people tell me they want to vermicompost or compost a large amount of organic materials, I urge them to first look into their state's regulations for these activities. Most US states have enacted composting regulations to protect public welfare and the environment. Regulations vary considerably from state to state, although most of them establish setback distances for organic materials recycling operations from dwellings, ground and surface waters, and property lines. They usually require leachate and runoff management plans, minimum temperatures that must be reached to eliminate pathogens, plans for reducing the risk of attracting rodents, insects, and birds (called Vector Attraction Reduction), and minimization of odors and dust.

Under certain circumstances, some state officials consider vermicomposting to be an agricultural activity to raise livestock (earthworms). If state regulators decide that your plans fit into this category, you will likely not need a composting permit.

State regulators may decide that your planned vermicomposting operation is similar to the activities of thermophilic composting facilities, and they may require you to obtain a composting permit. Although most state composting regulations do not mention vermicomposting, if you are planning a large vermicomposting operation that processes waste materials, or you plan to produce and sell vermicast, you will likely be required to obtain a composting permit. Composting regulators are employees of a state's environmental protection agency, usually in the solid waste management division. For example, in North Carolina the officials who regulate vermicomposting operations are in the Department of Environmental Quality, Division of Waste Management, Solid Waste Section. North Carolina offers the option of obtaining a demonstration permit for people who are just getting started; it is much quicker and easier to obtain a demonstration permit than a full permit. This allows folks to try out their idea and see how it works. After a year or two, they can apply for a full permit and expand their operation.

To find the composting regulators in your state, google "[your state] composting regulations." In addition, the US Composting Council has information about state regulations and contact information for each state (see the resource section at the back of this book). If you plan to sell vermicast, you should also contact your state's department of agriculture (DOA) to see if you need a facility license or product registration. The department may require you to submit samples of your vermicast for laboratory testing. (For more about DOA testing for vermicast and product labeling, see chapter 9, page 168.)

It's critical to comply with state-level regulations, but don't stop there. Check for local regulations (county, city, borough, et cetera) as well, including regulations related to land use and zoning, storm-water management, solid waste management, construction and occupancy, and public health.

Some regulations and permits are under county jurisdiction, and others are controlled by the city where your business is to be located. City and county governments often provide information on their websites about permits and requirements for starting a business in their area. To find your city's requirements for new businesses, you can type in "[your city name] regulations for businesses" on an internet browser. You can do a similar search online for your county government's laws for businesses by searching the term "[your county name] regulations for businesses."

If you plan to vermicompost food waste onsite at a restaurant, school, grocery store, office building, or other site, you will need buy-in from the local health department. They regulate businesses and institutions that generate food waste to protect public welfare. The health department requires you to ensure that the site is not infested with rats and insects, and some inspectors may object to a vermicomposting system that contains earthworms and other creepy-crawlies. Before contacting the health department, see if there are others in your area who are vermicomposting food waste onsite and ask them how they obtained approval from the health department. If the health inspector starts to object to your plans, you can describe what others are doing with their health department's blessing and provide their contact information.

It may sound intimidating to contact state and local officials about regulations and permits, and you may be tempted to not notify them of your plans. Many folks live by the adage, "It's better to beg for forgiveness than to ask for permission." However, state and local officials will not be forgiving if they discover what you are doing without a permit, and they can impose huge daily fines or shut down your operation and confiscate your equipment.

People often ask me to interpret how the state composting regulations will apply to their situation. I decline and tell them that my opinion doesn't matter. The state regulators are the only ones who can interpret and enforce the regulations.

Contacting regulators before you proceed will probably save you time and money. You may have a piece of land in mind to buy but the zoning regulations there might not allow vermicomposting. Or you may already own land and know right where you want to set up a composting and vermicomposting operation. Before moving forward, ask the regulator to visit with you at your property. The regulator will be able to tell you where the state and local governments will allow you to set up shop in keeping with regulations. For example, you may want to carry out your activities on the southwest corner of your property. The regulator may say, "No, that won't work because it's too close to property lines" (or a stream or well; or the land has too much of a slope). The regulator will also be able to help you determine which parts of your property would be in compliance with regulations.

So go ahead, contact your state compost regulator. If you are polite, honest, and cooperative, you will get the best results.

Prepare a Business Plan

It is important to develop a business plan that will guide you through setting up, managing, and growing your business. Business planning takes time and thought, but there is no substitute for a good plan.

Explaining all the details of writing a business plan is beyond the scope of this book. If I were starting a business, the first place I would seek out information is the Small Business Administration (SBA; see the resources section). The SBA is an independent agency of the US federal government that helps folks who have (or want to start) a small business. The agency offers loads of free information and services on many aspects of business planning, including:

- Choosing a business structure
- Writing a business plan
- Creating a market research and competitive analysis
- Calculating start-up costs
- Funding a business
- Choosing and registering a business name
- Obtaining federal and state tax ID numbers
- Obtaining a business bank account
- Applying for permits
- Obtaining business insurance
- Paying taxes
- Buying equipment and assets
- Conforming to legal requirements
- Hiring and supervising employees
- Marketing and selling products
- Accounting
- Exporting products

One fundamental decision is whether you will launch a large-scale commercial vermi-operation or stick with a small-scale business model. For a small-scale business, you (alone or with a partner) may bring in enough revenue to pay for your operating expenses and a small salary. At this scale you would likely sell primary and secondary products to local markets. Starting a large-scale operation requires more resources than starting small. Your business plan will help you determine the potential profitability of your endeavors. Your profits will depend on the revenues you receive minus the costs of capital expenses, labor, and other inputs.

A business plan for a vermicomposting business should include all of the following:

- List of the company's initial needs
- Description of products and services
- Description of how products will satisfy marketplace needs
- Size of potential market
- Product pricing
- Structure of the business
- Start-up outlays
- Capital and operation expenditures
- Equipment needed, costs, and life expectancy
- Feedstocks available and their costs or revenues
- Costs for insurance premiums, mortgage or rent, and labor
- Utility needs and rates
- Feedstock contaminants handling plan
- Operations plan
- Preventive maintenance program for all equipment
- Monthly expenses and projected revenues
- Financing plan (loans, grants, investors)
- Break-even price point
- Marketing plan
- Neighbor relations plan
- Production schedule to meet seasonal demands
- Methods for measuring product characteristics
- Methods for testing and screening vermicast
- Fuel delivery and containment plan
- OSHA compliance plan
- Contingency plans
- Timeline
- Executive summary

Your business plan will be an extensive document, the result of many hours of work.

Neighbor Relations and Contingency Plans

People often ask me what I mean by a neighbor relations plan and contingency plans. I tell people that in terms of neighbor relations, it's smart to be proactive. Contact all neighbors who live or are doing business adjacent to your site before start-up and educate them about what a vermicomposting operation is and why you are starting the business. Continue to keep them informed as you develop the site. It's ideal if your site is already screened from view by a fence or trees and shrubs, but if it isn't, planting a vegetation buffer is important. Out of sight is out of mind for those who may not be thrilled about living next to a vermicomposting operation.

Your contingency plans are one of the most important features of your business plan. What are contingency plans? They are your fallback plan in case some aspect of your business doesn't play out the way you hoped or something unexpected arises. You need a contingency for every aspect of running your business and marketing your products. Consider, for example, what you would do if one of these situations arises:

- Your feedstock source dries up.
- Feedstock that is delivered is contaminated with unwanted materials.
- Your outdoor vermi-system no longer works because of weather conditions (too hot, cold, wet, or dry).
- Neighbors complain about odors and pests.
- Equipment fails.
- The site is destroyed by a natural disaster.
- Your source of funding dwindles or disappears.

Trying to develop backup plans when you are in crisis mode rarely works, so do it while you are writing your business plan. For example, equipment failure is pretty much inevitable for any business, and it could be a huge setback unless you have a plan in place to deal with it. Your plan could include a list of area mechanics who could fix the equipment and establishments that could rent or loan you machinery until yours is repaired. Your business plan will never be a finished document. Once you get your operation up and running, you should scrutinize your progress periodically and make adjustments to your business plan as needed.

More Business Start-Up Details

Your business will need a name (and naming your business is one of the fun parts of business planning). After you decide on one, you should register it to protect it. There are four ways to register a name: entity name, trademark, domain name, and doing business as (DBA). It's a good idea to do all of them, because they serve different purposes. You don't have to use the same name for each type of registration. Entity name safeguards you at the state level. Trademark protects you at the federal level from others using your name. Domain name protects your business website address (URL) and your brand online. DBA doesn't give legal protection, but it may be required by your state, city, or county. See the resources section for information about choosing and registering your business name.

Another task is to register your business with your city or county, your state, and the federal government, depending on where your business will be located and its type. You will need a federal tax ID number, and you may be required to register with the secretary of state's office.

Deciding what type of business structure best suits your business requires research and reflection. These are your choices:

- Sole proprietorship
- Partnership (limited partnership LP) or limited liability partnership (LLP)
- Limited liability company (LLC)
- Corporation (C corp, S corp, B corp, close corporation, nonprofit corp)
- Cooperative association
- Hybrid ideal (nonprofit structure with for-profit business model)

All of the choices have pros and cons; only you can decide which structure will work best for what you want to do. Many vermi-bizzes choose the limited liability company model. Each model has diverse legal and tax consequences, so it would be wise to discuss the options with a tax attorney or certified public accountant.

As soon as you start spending money on your business or bringing in revenues, you should open a business account at a bank. You need to obtain a federal Employer Identification Number (EIN) first, so keep that in mind. Keeping your personal funds separate from your business money is a good idea, and a business account offers limited personal liability protection. If you need to make sizable start-up purchases, you may want to open a credit card account for your business, too.

Creating a site plan should be part of your business plan. Determine where to set up components such as vermicomposting bins and areas for feedstock receiving and processing, pre-composting, equipment storage and maintenance, harvesting and screening, packaging, and storage. Be sure to separate feedstocks and finished products. As you design your site, think about traffic flow, storm-water management, parking, and aesthetics for keeping your site

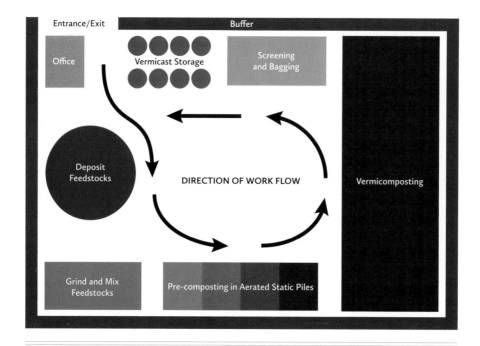

FIGURE 4.1. This vermicomposting operation site plan shows a counterclockwise flow for efficient processing of materials.

attractive and not messy. You will also need bathroom facilities onsite and perhaps an office.

Develop a Marketing Plan

Worm farmers often overlook the importance of marketing. I can't emphasize its importance enough. Before you start constructing your vermi-operation, figure out to *whom* you are going to sell and *how* you are going to sell to them, and then design your system to match your marketing plan.

Many people are excited about setting up their system, and that's natural, but ultimately the goal is to *sell* vermicast or worms. I can't tell you how many people have called me to ask a question like, "I just harvested 2 tons of vermi-compost—how do I sell it?" They gave no advance thought to marketing, and now they are panicked—with good reason.

A couple of decades ago, I visited a large family farm that raised a variety of crops. The farmer had decided to vermicompost the dairy manure produced on his neighbor's farm. He set up a large vermicomposting windrow operation and had a series of conveyors that would transport harvested vermicast from the field to a barn, where the product was automatically bagged. The farm had thousands of cubic-foot bags of vermicast stacked almost to the ceiling. Sadly, they never sold a single bag. What went wrong? Connect the dots in the picture I just described. It was a large farm run by family members who had a variety of tasks they had to accomplish every day. They possessed skills to set up an efficient, mechanized system that was highly effective. However, the family members were already too busy with their farming activities to focus on marketing. Or perhaps they may have lacked sufficient sales skills to sell their product, so the whole operation collapsed. If they had developed a marketing plan before moving forward with operation development, things would likely have turned out differently. As they wrote the plan, they might have realized their marketing limitations and decided that they needed to hire someone to sell their products on their behalf.

I hope this example helps you understand how imperative it is to develop a marketing plan before you move forward with setting up a commercial vermicomposting operation. Begin by conducting market research to quantify the market and ascertain who your competitors are. Marketing research needs to be ongoing and adjusted as needed. You can conduct market research by attending trade shows such as those sponsored by the US Composting Council; or reading *BioCycle* magazine; or visiting conferences about soil amendments, gardening, or landscaping. People are a great resource—you can interview other vermicomposters or talk with your cooperative extension agent. Reading

FIGURE 4.2. Be creative with packaging, both to boost customer interest and to keep packaging costs low. Worm Endings Unlimited sells vermicast in snack bags for $1. Worm Endings worms are sold in reused milk cartons covered with a nylon stocking.

is another important part of market research. Browse the internet for articles on marketing research or check out books such as *The Practical Guide to Compost Marketing and Sales* by Ron Alexander. If you feel that you want one-on-one assistance, contact a vermicompost and compost marketing consultant and hire them to either do research for you or coach you on how to do it.

Figure out what your local market may want or need. Check store shelves in big-box stores, nurseries, agricultural or feed stores, and garden shops to see what soil amendments they stock. Are they selling vermicast, or soil mixes with vermicast as an ingredient? Take notes regarding the pricing, quantity of material in the package, whether the package lists the percentage content of nitrogen, phosphorus, and potassium (NPK), whether the product is OMRI certified (this means that the product is acceptable for use by certified organic growers), and additional ingredients. Take photos of the front and back of the bag. What do you like about the bag design, and what would you do differently?

Zachary Brooks of the Arizona Worm Farm in Phoenix established a business model where he gets paid to take others' waste (landscape waste, vegetable waste, and horse manure), which he turns into value-added products. Zach sells vermicast in bulk and bagged, and he also sells a vermicomposting mix for gardeners. He will deliver product to customers (he charges a delivery fee), or they can pick it up at the farm. (For more about Zach's operation, see the "Arizona Worm Farm" profile in chapter 10 on page 197.)

Another part of marketing research is to determine what barriers or stigmas exist and figure out how to overcome them. What do I mean by a stigma? Here's an example. In eastern North Carolina there are a huge number of hog farms. For those living in houses nearby, the smell from the farms can be quite strong and unpleasant. Hog manure is flushed out of the buildings that house the animals, and the liquid and manure are collected in big holes in the ground called lagoons. During heavy rain events, such as hurricanes, the lagoons may overflow and contaminate waterways. Therefore, some people living in these areas have formed the opinion that pigs are smelly and cause problems for the environment. To them, pig manure is bad, and they would most likely not purchase vermicast or compost made from pig manure. Many people have a more favorable impression of cows (which is not necessarily justified) and consider them cleaner and cuter than pigs. Cow manure is a popular soil amendment, whether it is raw or composted. Given the choice, there is a higher probability that people would purchase vermicast made from cow manure than pig poop. It is unfortunate that this stigma would prevail—in many research studies vermicast made from hog manure had a greater positive effect on plants than cow manure vermicast.

Once you've completed your initial research, it's time to write your plan. Here are some of the questions your marketing plan should address:

What do you want to sell? Possibilities include worms, vermicast, soil mixes, worm bins, books, feed, bedding, worm bin accessories, soil amendments, videos, T-shirts, hats, and more.

Whom do you want to sell your products to? Possible customers include gardeners, other vermicomposters, landscapers, internet customers, big-box stores, wholesale nurseries, greenhouses, golf courses, resorts, lawn-care companies, vineyards, organic farmers, soil blenders, and cannabis growers.

What geographic location do you want to cover? Within 50 miles (81 km), nearby states, the whole country, or internationally?

When will you sell products? Spring only, spring and summer, spring and fall, or year-round?

What specific quantities will you offer? You can sell vermicast by the cubic yard or cubic foot or small volumes in bags or containers.

How do you want to package and label your products for sale? As a fertilizer, soil conditioner, potting mix component, topdressing, planter mix, tea, or extract?

What prices do you plan to set, and how much revenue do you anticipate generating?

How will you stimulate repeat sales?

What is your estimated total demand for vermicast in your target market? How much of that will you be able to meet with the estimated production from your operation?

What are the current vermicast or compost demand and purchasing cycles in your target market?

Determining the pricing of your products and identifying how you want to sell them is a key part of developing a marketing plan. The options are quite varied, so take the time to think through these questions:

Will you do in-house marketing or hire a broker or distributor?

Do you plan to sell to industry professionals or through retail outlets?

Should you join a distribution network?

Will you sell your vermicast in bulk or as bagged products, or both?

Will you set up a website for direct internet sales or sell onsite at your business, or in stores, or all three?

Is it better to target a larger market with lower-value goods or a smaller market with higher-value products?

If you decide to work with distributors, you need to do some research and understand how they work. A common mistake is to get trapped by a distributor who demands exclusivity. You should only agree to this restriction if they guarantee a certain number of sales.

Here is an example of a two-tier distribution system. You sell your vermicast for $1 to $2 per pound to a distributor. They sell it for $2 to $4 to a retailer, and they sell it to the public for $4 to $8 per pound.

Selling bagged products requires a bigger commitment and more financial risk than selling in bulk. Although bags will have greater per-unit value and more buyers (internet sales and mass merchandisers), they will cost you more in processing and supply costs, as well as other logistics. Selling vermicast in bulk is an effective way to start, because it brings in money with little upfront investment required. After your business is more established and you've built up capital, you can invest in the supplies and equipment needed to make bagged products. Here is an added incentive for bagging: People think products in bags are higher quality than materials sold in bulk.

To make more money, you can sell to fewer numbers of users in smaller packages. One vermicompost producer I know sells vermicast for $400 per cubic yard (0.8 m³) in bulk. She also sells cute little packages of vermicast for about $15, and homeowners readily pay for them. What they don't realize, however, is that they are paying $1,800 per cubic yard for the same vermicast as the bulk buyers are getting!

Additional sales can be made from offshoot ("derivative") products. Here are some options:

Sell mixes such as 90 percent compost with 10 percent vermicast (or 80/20).
Sell specialty mixes for specific types of plants such as flowers, vegetables, and so on.
Sell bulk mixes onsite according to customers' recipes.
Sell individual packaged products such as bat guano, greensand, mycorrhiza, rock dust, et cetera.

To get more ideas for offshoot products, check out some websites. The Worm Farm (www.thewormfarm.net) sells about 80 different products and mixes materials on the spot according to customers' wishes. Kellogg Garden Products (www.kellogggarden.com/products) sells a vast array of products targeted specifically for vegetable gardens, flower beds, lawns, potting mixes, cacti and succulents, acid-loving plants, raised beds, you name it—there's a product to meet the needs of every customer. Click around Kellogg Garden Products' online product list and you will see attractive packages that list the ingredients, what the mix may be used for, how much to apply, and how it will benefit plants and soil.

Market development takes an investment on your part—of money, effort, and time. It will take a great deal of patience because you will have to develop relationships with potential customers to convince them to trust that you have the best product for their needs. It can take two or three years to establish a loyal customer.

Education as a Component of Marketing

People in your area may not be familiar with vermicast. Therefore, education must be a major component of your sales and marketing efforts. Start with yourself: Do you have a good working knowledge about vermicast and how it can benefit plants and soil in your target markets? Once you have educated yourself, develop your educational marketing plan, which may include creating educational materials that accompany your products; adding educational information about your product into your sales pitch; speaking at cooperative extension meetings, lunch 'n' learns, garden meetings, conferences, service clubs, and workshops; and offering classes at your business.

Vermicast producers who offer educational seminars about their products at their business location report that about three-quarters of the attendees purchase other products while they are there, including worms, vermicast, worm bins, soil amendments, T-shirts and hats, books, and videos. When you teach classes, customers will see you as an authority, and your products will seem more credible. Customers will become loyal to your brand.

There are plenty of other fun and effective ways to educate customers. Try some of these ideas:

- Set up a page on your website explaining the benefits and uses of your products.
- Create a Facebook page, and regularly add photos of your operation and updates about your products. Post pictures of plants you have grown using your vermicast compared with those that were not grown with your product.
- Start a Twitter account for your business, and tweet about your products.
- Make brief YouTube videos.
- Write a monthly or quarterly newsletter about your products and end-user successes.
- Create a blog on your website, and post regular updates.
- Construct an educational display that you can set up at events, fairs, farmers markets, workshops, or trade shows.
- Write or be interviewed for articles in the local newspaper.

- Provide tours of your facility (some worm growers charge $1,000 per person or more for tours).
- Get to know garden writers in your area, and educate them about the benefits of your vermicast so that, in turn, they can write an education blog about it for their followers or an article for a newspaper or magazine.

Create a Promotional Plan

You're probably already daydreaming about your company name. Have you also thought of a logo? You need something eye-catching and unique that stays in folks' minds. Give some thought about how catchy you want it to be and what you want to convey about your company. Do you want a cartoon worm on your logo or something that conveys that you are a scientific authority?

Some producers claim that displays at trade shows are the best money they have spent to promote their company. Do your research first by simply attending some conferences to see how many people attend and how much action there is at various displays. This helps to determine which trade shows might provide the best bang for your buck. Hundreds of companies exhibit at the US Composting Council's annual conference (see the resources section). Your state may have a composting council or recycling association that you could join, and they often have an annual conference. You could also join or exhibit at conferences held by your target market, such as people involved in landscaping, horticulture, agriculture, turf grass, viticulture, and so on.

Setting up a website will give you the widest exposure for your business. It will probably attract visitors from all over the world. On a website you can share the story of how you got started, your mission and goals, how far your business has come, and the things you want to do in the future. You can show photos of your products, describe their benefits, and provide an easy method of payment for your customers. People are more likely to purchase your products if they can see photos of what you have to sell. If you include photos of you, your family, and your workers, it will build trust with your customers. They also like to be able to communicate with you, so provide an email address, phone number, and physical or postal address. You will build your reputation and elicit trust from your online visitors if you share more information. Browse the worm businesses featured throughout this book to find the URLs for their websites. You can learn a lot by seeing how others have set up their business websites.

You have several options for setting up a website. There are free website host sites, such as Yola and Weebly, or you can pay a yearly fee to register a domain name and pay a web hosting service to make your site accessible on the internet.

To create website pages, you can do it yourself or pay a professional web designer. Either way, you will have to write all of the content. You may be leaning toward paying someone else to design your website because it seems intimidating to do it yourself. However, you then also have to pay them to maintain your website and make any changes to it that you request. It can be quite costly, and if they go out of business, you may not be able to get your website back from them. The good news is that there are free website templates on the internet, and many are user friendly. Once you get the hang of it, it is easy to update and maintain your own website. See the resources section for more on website templates and online products to help with marketing and promotion.

Nowadays, when you want to find a restaurant, a gas station, or another business located nearby, you simply type the word into Google search, and it provides the option "near me" (say, "restaurants near me"). Consequently, it will help people be aware of your business if you put your street address on the header of your website, because it will appear on every page. Mentioning your worm farm's location in blog posts will also be detected by Google and added to Google Maps.

Think about the product literature you will develop, such as brochures and flyers. Will you advertise in print ads, radio, the internet, or television? When customers call, always ask, "Where did you hear about us?" Keep records of their responses so you can determine the effectiveness of your advertising.

Whom do you want to reach with your advertising—the general public? Green industry professionals? Landscape architects? Organic farmers? You could offer to give presentations for garden clubs, green industry lunch 'n' learns, service clubs, cooperative extension buildings, garden centers, nurseries, arboretums, schools, farmers markets, and libraries.

What to Avoid: Why Operations Fail

Not to discourage you, but I must share this disheartening news: Many large commercial vermicomposting operations in the United States have had to shut down. The failures happened for a number of reasons, but many times it had to do with rushing into business without enough knowledge, experience, and preparation. The primary reasons for failed ventures are listed below.

Lack of Knowledge

Learning about earthworm husbandry doesn't mean just reading about it. You have to experience in a hands-on way the practical details of how to care for earthworms. Start with 1 pound of worms, and after a number of months of keeping them alive and helping them to thrive, buy 5 pounds of worms. You will see that caring for 5,000 worms is different from raising 1,000. Then scale up to 10 pounds of worms. The point is, you need to scale up slowly. If you have never cared for a worm, you will be shocked if you impulsively buy 10 pounds or more of worms only to learn that you have no idea what to do with them. It will be an expensive lesson if they all die.

Starting Too Big

During the past 25 years, thousands of people have contacted me with questions about vermicomposting. Someone called me recently to discuss whether to buy a special type of worm bin called a continuous flow reactor system costing $50,000 that he had seen only in pictures on the internet. As our conversation progressed, I realized that he had very little experience with vermicomposting. He was asking me very basic questions—information that he should have figured out long before he even considered spending $50K for a worm bin.

I could tell you a hundred more stories about people who magically assume that if they buy the right type of system, everything else will fall into place. They have very little experience and have done almost no research, yet they're willing to invest a large amount of money on faith alone. I'll say it again: Determine your markets first, then master earthworm husbandry (if you haven't already done so)—and only then start shopping for the right piece of equipment for your planned operation. There are hundreds of details to work out before you buy equipment—and that's the reason for writing a detailed business and marketing plan.

Undercapitalization

If you've seen videos about large vermicomposting operations, you may feel inspired to develop a similar facility. However, to start a large worm farm, you will need $5 to $10 million. Do you have access to millions of dollars? If not, you'll probably need to reconsider the size of the operation that you want to establish. You could seek investors, but it can be very difficult to convince them to invest, and if you're not careful they could end up owning more of the business than you do.

You need to be sure to account for full production costs, including utilities, equipment, worms, feedstocks, maintenance, labor, taxes, monitoring, testing, marketing/promotion, packaging, shipping, and pest control. And then add in a rainy-day fund to cover the costs of the unexpected, such as an accident that results in damage to equipment or an injury.

People frequently ask me if there are grants they can apply for to fund their business venture. They are fired up about how vermicomposting is going to save the planet and assume that the government is willing to pay for their revolutionary vermi-concept. Sadly, that is not the case. If you don't have lots of cash on hand, you have probably noticed that times are tight and there is less money available for salaries and services. The government is tightening its belt and funding fewer projects.

I'm not trying to discourage you from seeking to obtain government grants. Federal grants are more competitive than state or local grants. One excellent resource about grant funding is www.grants.gov, a website where you can access 900 grant programs obtainable from 26 federal grant-making agencies. At this website you can investigate whether you may be eligible and learn about available funding and how to apply.

Taking out a loan is another way that many people fund their business start-up. It takes time and effort to research and apply for loans, but it can be quite productive. See the resources section for links to loans through the USDA.

Difficulties with Regulators

If you shrugged off my advice in the early section of this chapter about state and local regulations, go back and read it again. Regulators can be your friend or your foe, so tread lightly. If you are impolite, arrogant, or dishonest with a regulator, you may end up regretting your actions. Remember, they have the power to shut down your business. They can seize all of your equipment, including your computers.

Regulators are not mean people who wish to make your life miserable. They are carrying out the carefully crafted rules that are designed to protect people, plants, and the environment. If you follow the rules, it will benefit your business. You will produce a safer product that will, I hope, increase your sales.

Depending on your state's composting regulations, you can get into trouble if you:

- Fail to report a fire at your site
- Have odor complaints
- Exceed permitted dimensions of piles or units
- Fail product testing standards or do not test your product according to the required schedule
- Have storm-water or runoff issues, such as puddling and ponding on your site or visible runoff
- Fail to update or have inadequate financial assurance for closure (this is part of state regulations and may or may not apply to your operation)

- Allow piles to become contaminated with plastic or other foreign objects
- Don't maintain buffer zones, causing off-site litter and dust
- Fail to maintain required operational records, submit quarterly or annual reports late, or the like
- Do not have a certified compost operator onsite (if your state requires this for composting activities)

Odors are the number one problem that causes composting operations to be shut down by regulators. That may not be the case for vermicomposting facilities, but the possibility is there. Odors can be caused by raw feedstocks, compost or worm bin mismanagement, and leachate or runoff. Manure, whether from animals or humans, can be quite smelly, and so can food waste. Once these stinky feedstocks are added to the worm bin, the smell can dissipate within 24 to 48 hours if you are managing the bin correctly.

Another reason vermi-facilities get shut down is due to complaints from neighbors. Most often the complaints are about odors, but there can be additional issues. Other activities that attract complaints are high traffic volumes to and from your site, loud machinery, excessive dust, and a junky appearance.

Composting operations have been shut down for violating the Clean Water Act. This has happened when rainfall that hit raw feedstocks or compost piles ran off into streams or when rainwater catchment ponds overflowed and contaminated creeks.

Inadequate Marketing Methods

If you want your business to be a success, it is essential that you understand the benefits of your vermicast product. Don't copy and paste what others have posted about their products on their websites. Your product has unique attributes, and it is up to you to discover what they are. You can do small-scale usage trials with cooperative extension centers or end users, or apply for grants to conduct research with a university. Some of the university departments that may be interested in using your product for plant trials include horticultural science, plant pathology, crop science, entomology, and agroecology.

You must be able to differentiate your vermicast from compost. Most people have heard of compost and have general impressions about how it improves soil and plant health. However, huge numbers of people have never heard of vermicast or worm castings and have no idea what it is for or how to use it.

You lose credibility if you talk about "the magic of vermicast" but do not cite authentic research. The bibliography at the back of this book includes listings of excellent articles describing scientific trials that have been conducted with vermicast. It would be wise to read these studies. Many of them can be downloaded for free from the internet. Just type the name of the study you want to read into your internet browser.

Ponzi or Pyramid Schemes

For decades the vermicomposting industry has been plagued by Ponzi schemes. Charismatic individuals will lure investors into buyback contracts costing $10,000 to $100,000. In the late 1990s a company called B&B Farms created a Ponzi scheme in the United States that scammed more than 2,400 people in 43 states who wanted to get into worm farming. Investors were promised they could obtain a sizable income with little effort. For their payment of tens of thousands of dollars, they would receive worms and assistance for raising them. Then they could sell the worms back to B&B Farms for up to $9 per pound. To many, it seemed like an easy way to make money with a guaranteed market. However, the worms that growers sold to B&B Farms were not being utilized for end-use projects; instead, they were just being sold to new growers, which is illegal. The Department of Securities and attorneys general in several states started clamping down on B&B Farms, so the company declared bankruptcy and shut down operations. Some investors had not set up any contingency strategy for how to sell their worms to anyone other than B&B, and they were left in a tough situation. People lost their investments and were stuck with worms and no market. Some people had mortgaged their houses, used funds set aside for their children's college costs, or spent their life savings on this pyramid scheme.

That scenario has been repeated numerous times throughout the United States. I always try to stress to those who attend my presentations or ask for business advice, "If it sounds too good to be true . . . it probably is."

Insufficient Insurance for Facility and Products

All sorts of unexpected incidents can cripple or shut down your business. For instance, there are many ways that employees could get hurt or become ill at your vermi-operation. Whether it is their fault or yours, you will probably be liable. Therefore, be sure to have a workers' compensation policy that covers medical care and lost income for employees, death benefits for their families, and your possible liability for claims. In addition, you will need (or be required) to have unemployment and disability insurance.

Property insurance covers risks associated with your site, building, equipment, inventory, tools, and computers. If these things get damaged or destroyed, your policy may cover your losses.

Business liability insurance pays for medical costs if someone gets hurt on your property, and for property damages or injuries that are caused by you

Insurance Options

Other types of insurance that may offer useful coverage for your business include:

Legal expenses insurance
Fire insurance
Flood insurance
Combined liability insurance
Rehabilitation insurance

Material damage insurance
Stock and business interruption
    insurance
Waste liability insurance
Plant and machinery insurance

or your employees. If you are sued, this insurance will help cover your legal defense costs and any money you are required to pay if you lose the lawsuit, such as punitive or compensatory damages. This insurance can also safeguard you from liability if you are renting a property and you cause damage or there is a fire. Business liability insurance can cover your expenses, too, if you are sued for slander, libel, or copyright infringement.

This may sound surprising, but if you want to sell your product to the greenhouse industry you should carry $2 million or more in product liability insurance. One worm farmer had to take out a $1,000-an-hour insurance policy when a large greenhouse grower wanted to try using his vermicast in his automated plug production line. If the vermicast is sticky, it can plug up the equipment and shut down their assembly line. You will be blamed for that, and your contract will require that you pay for their losses.

"Putting Their Eggs in One Basket"

Some worm farms and other businesses have had to close their doors because they relied on a majority of their income from one client. I know someone whose family business had been handed down through generations. He landed a contract with a major retailer and built his business around it. I toured his operation and was amazed by all the sophisticated equipment he had purchased. You have probably figured out where this story is going . . . a competitor came along who offered lower prices, and the retailer switched all its business to the competition. My friend had substantial debt due to all the equipment purchases, and now most of his market had disappeared. Sadly, he had to close the doors of his family business.

A wise rule of thumb in business is to never put more than 20 or 30 percent of your business in one customer's hands. Diversify your market by having numerous customers and a variety of products to sell. The Worm Farm in Durham, California, has a very diverse market. They sell a large array of products, and their business plan has changed over the years to accommodate customer demands. They don't have just one large customer; instead, they have hundreds of them, including growers and landscapers. Their customers know they can rely on The Worm Farm to provide what they need when they want it.

CHAPTER 5

Designing Your Vermicomposting Setup

A s you talk to other worm farmers or read about their operations, you will learn that there are many ways to create a vermicomposting system. Some folks do it outdoors, while others set up their operations indoors due to the climate or other concerns.

The vermi-system where the worms live and work is the heart of any operation. The most common types of vermi-systems include windrows, wedges, pits, trenches, beds, bins, and continuous flow-through bins (CFTs), which are also known as CFT reactors. For bins and CFTs, worm farmers can choose to buy manufactured units or build bins themselves from wood or metal, often by repurposing items or materials that are readily available.

In this chapter I discuss factors to weigh when you are choosing what type of vermi-system will be best for your setup. Next I help you think through space requirements for different aspects of your operation. Finally, I provide descriptions of the most common types of systems and their advantages and disadvantages.

Parameters for Choosing a System

When deciding whether to set up a vermicomposting operation indoors or outside, and which system to use, take into account your climate characteristics, available onsite area, quantity of feedstocks, available funding, existing buildings, labor (needs, costs, and availability), state composting regulations, local zoning and health rules, and predators. Let's consider each of these factors in turn.

Climate

Local climate is one of the biggest limiting factors in your choices of vermicomposting system. This is because earthworms are sensitive to fluctuating temperatures, and as described in chapter 3, *Eisenia fetida* operates best at 60 to 80°F (16–27°C). If you live in a mild climate that doesn't often get too hot or too cold, you can set up an operation outdoors and use windrows, which is the least expensive worm farming option. Because of its temperate climate and limited rainfall, California has the most outdoor worm windrows of any state, and many vermi-operations in California use continuous flow-through bins with a simple roof overhead to shield from the sun and rain. They have the advantage of no heating bills!

Many years ago I visited a farm in Texas that had 7 acres (2.8 ha) of worm windrows. Although the area reached over 100°F (38°C) on several days during the summer and experienced freezing temperatures in the winter, they successfully managed the windrows by using moisture to counteract extreme temperatures. During the summer, they lightly misted water on the windrows in the evening when temperatures were cooler. When temperatures dropped below freezing during winter, they turned on their windrow irrigation system so the water would form an insulating protective layer of ice on top of the pile.

In most cases, however, if you live in an area of the US with wide temperature variations, you should establish your commercial worm farm indoors, where you can easily moderate the temperature. A profit-making venture cannot afford to have the worms slow down their consumption, excretion, and reproduction due to excessive or inadequate temperatures.

If you are producing vermicompost for use at an institution or school or by a municipality and are not concerned with optimal production, you can keep your worm bins outdoors. There are plenty of outdoor bin setups around the United States and Canada that work throughout the year, even in areas with heavy snowfall.

My worm bins at the Compost Learning Lab are inside a barn that is not fully enclosed and has no electricity. I'm not vermicomposting to achieve high production, but if that were my goal, I would keep my operation inside a building with better temperature control. Since my worms are subjected to wild temperature swings, they don't move as quickly as those employed for commercial production— it takes my worms a week to consume the dairy manure they are fed, whereas a for-profit operation would likely feed their worms two to three times per week.

Available Onsite Area

Vermicomposting is a shallow process, and a vermi-system can take up quite a bit of horizontal space. The size of your vermicomposting site will help you decide which type of system will work best. If you have limited square footage, you will probably want to use some type of bin system. If you have lots of land and live in a mild climate, consider windrows, because they tend to be the least expensive option.

Quantity of Feedstocks

The quantity of feedstocks you plan to process will influence the type of system you choose. If you are planning to process large quantities of waste, you will want the most efficient system possible, and that will likely mean a continuous flow-through reactor. The upfront costs for CFTs are higher than other systems, but in the long run they will save time by automating the vermicast harvest and by cutting down labor costs.

If you do decide to tackle a contract to process a large quantity of feedstocks, you would also be wise to pre-compost, thus reducing the volume of materials by half. This will save you the financial outlay of having to purchase or construct a larger vermi-reactor. (See chapter 7, page 124 for more information on pre-composting.)

Funding Available

A continuous flow-through bin is generally the most expensive vermicomposting system, so you need to determine if you can afford to buy one (or, if necessary, more than one). It's also worth considering that you can make your own for as little as one-fifth the purchase price of a new, commercially made reactor. (To get an idea of ballpark costs, browse through the profiles of worm farms and other large operations in chapter 10.)

Existing Buildings

Vermicomposting systems can be set up outdoors, or housed inside a structure such as a Quonset hut, greenhouse, pole barn, hoop house, warehouse, or farm building.

You have an advantage if there are existing buildings on the property where you intend to set up your vermi-operation. Old farm buildings that once housed chickens, cows, horses, or other livestock can be repurposed to house a vermi-system. They can work well for farmers who want to diversify their income stream, or by folks who choose to buy a farm property and go into vermicomposting after retiring from their non-agricultural jobs. Medium-sized businesses will sometimes set up bins in the basement or garage of an existing building.

FIGURE 5.1. The Worm Farm in California repurposed these old chicken houses for their vermicomposting operation.

Labor Needs, Costs, and Availability

Commercial operations need to determine how much labor will be required to run the business, and how to obtain it. If your operation is at a scale where you can manage the daily tasks by yourself, you may need only the occasional temporary worker at those times when you are harvesting your vermicast or performing other labor-intensive tasks. Some less expensive labor pools to draw from include migrant workers, people performing community service, correctional institution inmates, and temporary employment services. If labor is very expensive in your area, that could influence your choice of vermi-system. It may be critical to find the funding for a continuous flow-through system so that you can handle the work of the business primarily by yourself.

State Regulations and Local Rules

As discussed in chapter 4, most states have regulations that govern composting operations. But what do composting regulations have to do with your choice of a vermicomposting system? If your operation falls under state composting rules, the state regulator will have to approve the type of system you use, where you can locate it, and whether or not it needs to be under a roof. Local zoning and health rules may also apply to your site or choice of system. If you glossed over the discussion of regulations in chapter 4, turn back now and study them again.

Nuisances

Local predators can be a critical factor in determining whether to keep your operation indoors or outdoors. For instance, depending on where you live, bears may be a serious consideration if you are using human food waste or another type of feedstock that would be attractive to them, and you may have no other choice than to keep your vermicomposting system inside a secure building. Do research, and think about any nuisances that may disrupt your vermi-operation, and then take preemptive measures. For example, if you plan to have outdoor bins and moles could be a problem, you should install a protective barrier that will prevent them from accessing the worms.

Space Requirements

As you plan your vermi-operation, give careful thought to how much space will be needed for all aspects of what you plan to do. Here are some of the most common space requirements you may need to take into account:

FLOW OF MOVEMENT. It's important to have an efficient flow of movement at your site so you can minimize travel distances for handling materials. Retracing or crisscrossing paths as you move materials wastes both time and money. Leave enough space around all the components of your operation to prevent traffic holdups and avoid putting people in peril as they navigate the site.

FEEDSTOCK DELIVERY AND STORAGE. You will need a designated area to receive feedstocks and perhaps store them for extended periods of time. Depending on your state regulations, this area may need to be kept undercover.

FEEDSTOCK PREPARATION AREA. Depending on the size of your operation, you may need a separate area for chopping, grinding, or mixing feedstocks. Research equipment options and their footprints. Include space for a skid loader or other machinery to move things around.

PRE-COMPOSTING AREA. If you plan to pre-compost your feedstocks, decide what type of composting system you will use and figure out the dimensions. Locate it near the feedstock processing area, and maintain clear pathways for equipment to move the feedstocks to the compost bins and then transport the compost to your vermicomposting system. More information about pre-composting is in chapter 7 on page 124.

THE VERMI-SYSTEM. Depending on the type of vermi-system you have chosen, you will need plenty of space for it—up to 8 feet (2.4 m) of width for each system and whatever length you can accommodate. In addition, you will need room to move around the system, whether you are on foot or riding around on machinery, such as a forklift or front-end loader. Strive for a minimum of 3 feet (91 cm) of space between worm beds or bins. You will need this area to be clear for carrying out a variety of tasks, including feeding, checking your worm population, inspecting the bin, harvesting, watering, and cleaning.

BREEDING BINS. Even if your focus is on producing vermicast, you may want to breed worms in separate breeding bins. These additional worms can be used to expand your production operation, or replace worms that die or crawl away. You will need to allow space for continuing to add breeding bins over time, too, if you decide to increase the worm population.

STORAGE AND UTILITY AREAS. You will need space to store tools, equipment, hoses, cleaning supplies, and other paraphernalia. Some of this gear can be fragile or expensive, so you may want a closet or storeroom where you can lock everything up. Some type of toilet facility is essential, and you should also consider whether you want to have an office onsite.

HARVESTING AND PACKAGING AREA. An area for harvesting and screening vermicast is a must. In addition, you need space to store the vermicast, and perhaps package it for sale. If you plan to sell earthworms, you will want tables for counting and weighing worms, and a place to store packing and shipping supplies.

EXPANSION SPACE. As your vermi-operation gets under way, you may eventually decide to expand or adopt another practice that requires more room. Plan ahead for that, and designate some space for future growth.

If you are vermicomposting onsite at a business such as a restaurant, space may be unusually tight. Many restaurants have limited space, but you can explore the possibility of vermicomposting in the basement, on the roof, or in an alley.

An IKEA store in the Midwest implemented a creative solution for space limitations as part of a pilot program to vermicompost their cafeteria food waste. They hired a consultant, who explored options for locating worm bins. IKEA's loading dock would have been the best site, but they couldn't spare any space for worm bins, so their vermi-consultant decided to set up four continuous flow-through bins inside two box trucks that could be moved out of the way as deliveries arrived at the loading dock. The bins were about 20 feet (6.1 m) long but only about 3 feet (91 cm) wide and positioned flush against each side of the interior of the truck, with an aisle in between where the consultant could carry out earthworm husbandry tasks.

Types of Vermi-Systems

Let's take a look at the major types of vermi-systems that worm farmers are using to grow earthworms and produce vermicast. I start with the simplest—a windrow—and then cover pits, bins, and continuous flow-through bins. For more details on how to harvest vermicast from various systems, refer to chapter 9.

Windrows

Windrows are linear piles containing bedding and feedstocks that will eventually reach heights of 24 to 30 inches (61–76 cm). They are typically 4 to 8 feet (1.2–2.4 m) wide and however long the space allows. Many people living in milder climates, such as the West Coast of the United States, have established outdoor windrows for vermicast production, though windrows can also be set up inside large buildings.

A front-end loader, skid steer, shovel, or pitchfork can be used to make a windrow. Simply spread a 6- to 8-inch-deep (15–20 cm) layer of bedding of the desired width and length on bare ground and add 1 pound (0.5 kg) of *Eisenia fetida* earthworms per square foot (0.09 m²) of surface area. It's important to smother vegetation first so grass and weeds don't grow into the windrow, which would interfere with harvest. Since composting earthworms do not burrow, you don't have to worry about them migrating out of the windrow into the soil. After the worms have settled into the bedding, apply feedstock in a layer 1 to 1½ inches (2.5–3.8 cm) deep, and wait until the feed has been consumed before adding more. During cold weather, 3 to 4 inches (7.6–10.2 cm) of feedstock may be applied to the top of the windrow to help keep the worms warm. Although many worm growers have success with outdoor windrows, some folks report problems with excess moisture, anaerobic areas, and predators such as moles and birds. Vermicomposting in windrows is a slow process that can take up to 12 months to produce finished vermicast.

The Wedge System

Some people use a variation on windrows called the wedge system. As with normal windrows, this vermi-system can be used either outdoors or inside a structure. Those using the wedge system maximize space by applying feedstock in layers at a 45-degree angle against a finished windrow. Worms in the windrow will eventually migrate to the fresh feed. Keep adding layers of feedstock until the second pile reaches the depth of the first windrow, and then start a new wedge windrow. The worms will keep moving laterally through

FIGURE 5.2. This shade house at Arizona Worm Farm houses a dozen wedge-style vermicomposting windrows. *Photograph courtesy of Zachary Brooks.*

the windrows toward the new food. After three or four months, each wedge windrow can be harvested.

A wedge can be created from a pile instead of a windrow. For example, at Arizona Worm Farm, Zach Brooks starts a wedge by creating a 3-cubic-foot (0.08 m³) pile of pre-composted food waste, horse manure, mulch, and shredded cardboard at the corner of two low walls. He adds 10 pounds (4.5 kg) of worms to the pile. Once a week thereafter he adds 1 inch (2.5 cm) of the same pre-compost to the outer slope of the pile. He also adds a light layer of mulch over the compost because it helps keep the wedge cooler and makes the vermicast more fungal dominant. It takes four months for the wedge to "grow" 1 foot (30 cm) in length. When the wedge reaches 6 to 7 feet (1.8–2.1 m) long, Zach starts harvesting the wedge from the back end. Two hundred pounds (91 kg) total of feedstock is added to one wedge, and they harvest 100 pounds (45 kg) of vermicast. When I talked to Zach, the ambient temperature was 105°F (41°C), but in the center of his wedges the temperature was in the high 70s (about 25°C).

Pits or Trenches

Pits, or trenches, have been used for decades. As with other vermicomposting systems, you lay down moist bedding, add worms, and apply thin layers of the feedstock on top. People like pits for vermicomposting because they are simple

FIGURE 5.3. Earthworms consume hog manure in the dark-colored trenches inside this shelter at a hog farm in North Carolina. The farmer straddles a trench with a tractor-pulled manure spreader to dump food for the worms.

holes in the ground—no construction required. The earthen sides provide insulation against both hot and cold weather. Some people choose to line their pits with cement blocks. However, raising earthworms in pits can strain your back from bending over so often. Another disadvantage is that pits can be flooded by heavy rains.

A successful trench model was established in North Carolina in the early 2000s. A hog manure vermicomposter named Tom Christenberry was having trouble with his continuous flow-through bins overheating, so he created a trench system that allowed the ground to help moderate excessively cold or hot temperatures. Inside a pole barn he dug five trenches 4 feet wide by 200 feet long by 21 inches deep (1.2 × 61.0 × 0.5 m). Between the trenches were hard-packed dirt and gravel lanes that would support a tractor. To feed the worms, he drove a tractor straddling the top of a trench (see figure 5.3). The tractor pulled a manure spreader that was calibrated to deposit hog manure into the trench, and a worker would then use a rake to evenly spread the manure 1 inch (2.5 cm) deep across it. He carried out this feeding process weekly using raw hog manure that had been flushed out of a barn and run through a solids separator.

Before you commit to setting up a pit system, ask yourself whether you will be able to handle all of the bending that it will require to apply feedstock, check the worms and bed conditions, and harvest the vermicast. I used to think I could work with pit systems, but I now realize that I couldn't deal with getting on my knees, bending over, and reaching down into the pit on a regular basis.

Bins

Worm bins are a type of containment system that can be built, purchased, or repurposed. They are set up like the previous systems described—with moist bedding on the bottom, worms added on top, and then feedstock applied in thin layers on top of the bedding.

To make worm bins, worm growers often repurpose discarded items, including refrigerators, livestock tanks, tubs, barrels, wooden or plastic boxes, mortar trays, plastic buckets, washing machine tubs, and other containers. If such containers have solid bottoms or sides, drill holes for drainage and airflow.

Earthworm beds and bins can be built by hand from a variety of materials, such as lumber, plywood, concrete or cinder blocks, fiberglass, structural clay tile, poured concrete, or bricks. Wood is one of the most popular materials, but raw wood may decompose within 4 to 10 years, depending on the climate and other factors, so many folks will cover it with plastic or coat it with raw linseed oil. Avoid using redwood, cedar, or other aromatic woods to construct the bins; they contain tannic acid and resinous saps that are damaging to earthworms.

FIGURE 5.4. Here are two different styles of concrete-block worm bins: one at Camden Street Learning Garden in Raleigh, North Carolina (*top*), and the other at a farm in the Dominican Republic (*bottom*).

Worm farmers throughout the world build concrete-block bins, often on bare ground. They are inexpensive, easy to obtain, and won't rot. Concrete also provides more insulation than plywood, so it will keep your worm bed warmer in the winter and cooler in the summer.

FIGURE 5.5. This inexpensive vermicomposting shed was constructed of concrete blocks, shade cloth, tree limbs, and yagua leaves in the Dominican Republic.

FIGURE 5.6. Inside the shed in the Dominican Republic, worm beds lined with concrete blocks are on each side, with empty space in between for carrying out earthworm husbandry tasks.

Remember that birds love to eat worms, so if your bins aren't lidded you should use netting to keep out avian predators. Always keep direct sunlight off the worms, too, by either locating the bins in a well-shaded area, placing them under a roof, or covering them with lids. Figure 1.3 shows outdoor worm bins at a California school.

The worms in my Worm Barn at the Compost Learning Lab live in two MacroBins. These containers are manufactured by a company called Macro Plastics to store and transport vegetables. When I first saw one, I immediately thought, *This would make a good worm bin!* and decided I had to acquire some. The MacroBin is constructed of impact-grade copolymers and has smooth surfaces and ventilation slots on every side, as well as the bottom. It is approximately 45 inches long, 48 inches wide, and 34 inches high on the outside (1.1 × 1.2 × 0.9 m). The inside of the bin is approximately 41 inches by 45 inches by 29 inches (1.0 × 1.1 × 0.7 m). It can accommodate 1,300 pounds (590 kg) of materials and can be lifted by a forklift from any side. The MacroBin costs about $230, with discounted rates for bulk purchases. I wasn't aware of anyone else using these agricultural bulk containers as worm bins until I saw 10 of them set up for vermicomposting while touring a composting site in San Diego!

Garry Lipscomb and Bill Corey, the owners of NewSoil Vermiculture LLC in Durham, North Carolina,

FIGURE 5.7. These MacroBins for vermicomposting at a composting facility in San Diego are the same type I use at my worm barn in North Carolina.

made bins out of plastic 55-gallon (210 L) drums for the commercial operation in the basement of their home.

Many worm farmers repurpose plastic IBC totes (intermediate bulk containers) to make worm bins. They are large tanks used to store and transport 275 to 330 gallons (1,040–1,250 L) of liquid or bulk materials. IBC totes cost less than $150 new, with versions housed within metal cages available for approximately $100 more. These tanks are being repurposed for a variety of uses, so it's pretty easy to acquire them used at a lower price. See chapter 9 for information about harvesting vermicast from worm bins.

Batch Systems

In a batch system everything is added to a container at once—worms, food, and moisture—and the bucket or other container is then covered and left alone for 30 days. By then the worms will have consumed their food, and you will need to separate them from their castings.

I once visited a vermicomposting operation that had set up a batch system consisting of stacked plastic pots. A dozen plastic pots sat on a pallet, each filled halfway with feedstock and worms. A large piece of cardboard covered these pots, and 12 more pots were placed on the cardboard. This was repeated until the layers of pots were stacked six tall, for a total of 72 buckets on one pallet. This worm farmer had 35 pallets, so overall there were 2,520 buckets, each of which needed to be filled and emptied once a month.

The benefit of the batch system is that you can utilize vertical space efficiently by stacking the containers. However, it requires a substantial outlay to purchase the containers and equipment to move the pallets around. It is a labor-intensive system, and it is challenging to achieve and maintain optimal moisture levels inside the containers. In addition, the system requires a trommel or shaker screening system to separate the worms and the vermicast. These harvesting systems are covered in detail in chapter 9.

Continuous Flow-Through Bins

Continuous flow-through (CFT) bins are a type of raised bed system that is very popular for worm farming operations of any size. The open raised beds (built on legs) have a mesh bottom, also called a grate; this design provides air to both the top and bottom of the bin. With this system, you feed the top and harvest vermicast about two months later through the bottom.

Manufacturers of CFT systems are few, and many folks have found it more cost-effective to construct their own system, or pay a fabricator or welder to do

FIGURE 5.8. Benny Erez designed and built this continuous flow-through bin at Eco City Farms in Maryland.

the job. Some build their CFT bins with ⅝- or ¾-inch (16 or 19 mm) untreated plywood sides, while others make the whole bin out of galvanized steel. CFTs are often constructed using 5-foot (1.5 m) pallet racking with spacers. The beds range from 3 to 8 feet wide (0.9–2.4 m), with length dependent on the available space, and the bins are usually about 2 feet (61 cm) deep. You need space all the way around continuous flow-through bins—for people and equipment. For example, a 15-foot-wide (4.6 m) building can house two 5-foot-wide (1.5 m) bins while still allowing adequate space for movement.

Although most continuous flow-through bins have grate openings that are 4 inches by 2 inches (10 × 5 cm), some worm farmers install grates with 2-by-2-inch (5 × 5 cm) openings. The grate size for a CFT bin is dependent on the type of feedstock used, because material can cave in or fall through if the grates are improperly sized.

If you do not coat the interior of the plywood sides, they will decompose within four to eight years. To prevent this, paint the wood with waterborne sealer or food-safe wood finish. Do not treat the wood with oil-based sealants, because they contain petroleum distillates that could harm your worms.

After the bin is set up and ready to go, most people will apply two to four layers of newspaper or one sheet of brown packaging paper on the grate. Then they add 6 inches (15 cm) of moist bedding and introduce worms at a rate of about 1 pound (0.5 kg) per square foot (0.09 m²) of surface area of the bed space (length multiplied by width, not depth). For example, a CFT bin with a bed measuring 8 feet (2.4 m) long by 5 feet (1.5 m) wide would have 40 square feet (3.7 m²) of bed space.

To harvest vermicast off the bottom of a CFT bin, some type of manual or automated mechanism needs to be installed. Many worm farmers install a drive system with a cable that travels back and forth on top of the grate at the bottom of the bed. A T-bar on a continuous chain functions as a breaker bar that slides across the top of the grate, and folks will often use two U-bolts with clamps to connect the breaker bar to the cable. Clevis socket fittings are another option. The breaker bar rides back and forth on a cable and slices off 1 inch (2.5 cm) of castings.

Instead of attaching a breaker bar to a cable on a winch, the engineer who designed the CFT bins for Germina's vermi-operation in Turkey (see the profile in chapter 10 on page 200) installed roller chains to move the breaker bars. Picture a bicycle chain driven on a toothed wheel called a sprocket, except the chain is thicker and more durable.

Be careful about purchasing commercially manufactured CFT bins. You should be especially cautious with brand-new technology that hasn't had its bugs worked out, but even if a manufacturer has been making CFT reactors for a long time, it does not mean they are high-quality products that will perform well. Look very closely at the breaker bar system and the fasteners they use to connect the cable to the bar. Just like when you're purchasing a used car, it is wise to have a mechanic examine the equipment thoroughly to ensure it is well made and will hold up—ideally you should ask someone with welding or construction expertise. If you're not able to see the system in person, ask for detailed photos of the equipment and references from customers.

Stacked Bins on Pallet Racking

Some worm farmers maximize vertical space by stacking bins on pallet racking (steel shelves that can hold a lot of weight). When the city of Middleton, Connecticut, had a vermicomposting program, they purchased several 8-by-4-foot (2.4 × 1.2 m) worm bins and stacked them on separate pallet rack shelves.

Mass Wiggle in Petaluma, California, stacks their continuous flow-through bins three high to maximize vertical space in their warehouse. In total, they have 24 bins that are 70 feet long by 8 feet wide and 22 inches deep (21.3 × 2.4 × 0.6 m). Harvested vermicast drops onto the bins below, which are covered with erosion cloth. The workers climb ladders to access the bins at higher levels.

Covering the Bed

For all of the vermi-systems I have described, covers are optional. Having a lid or cover does provide several benefits, though, such as helping to hold in

FIGURE 5.9. These worm beds at Green Organic in Afghanistan are covered with burlap bags. *Photograph courtesy of Qasim Bakhshi.*

moisture and keep out light and guarding against the sun's heat and UV rays, rain, drying winds, and predators. Some growers choose to install permanent lids or covers, while others use them occasionally when needed (usually due to weather conditions). Covers used by worm farmers include burlap bags, old carpet, tarps, large tropical leaves, tin sheets, cardboard, paper egg or produce cartons, plywood, straw, and shade cloth.

Accessing the Worm Bed

Whatever type of vermi-system you decide on, it is vitally important to set it up so that you will be able to inspect every area of the worm bed. Even if things seem fine on the side of the bed that is easily visible to you, do not assume that everything is going well throughout the whole system. You need to be able to monitor the middle and the far side of the bed. Make this visual inspection by leaning over the worm bed—never walk on it! Keep this in mind while you are designing your system. Ideally, you should be able to access both sides of your worm bed and easily examine the middle. This is why many worm bins are only 4 or 5 feet wide (1.2–1.5 m): to ensure that you can easily bend over and see the middle of the bed. People who are taller and more flexible than I am are comfortable with widths up to 8 feet (2.4 m), but this would pose a challenge for me with my bad back.

FIGURE 5.10. A custom-made wooden lid protects this worm bin made from a section of culvert pipe at Apple Blossom School in California.

A stacked bin system maximizes use of vertical as well as horizontal space. However, like a pit system, this type of setup can be hard on the body, because it entails climbing ladders or shimmying up the supports to look in the bins. Plus it can be difficult to examine the entire bed with this type of setup. You should also be aware of continuous flow-through higher-than-normal bins that require taking a couple of steps up to look into the bins . . . then two steps down . . . move the ladder and repeat. That would get old in a hurry with 200-foot-long (61 m) bins.

A final issue to consider is accessing the underside of a CFT bin that has a breaker bar or some other type of automated harvesting mechanism on the bottom. If the cable or sprockets break, you will need room to crawl under the bed to fix it—while lying on your back with your arms upraised. What a nasty, uncomfortable job that would be!

CHAPTER 6

Vermicomposting Basics

In the previous chapter you learned how to decide what type of vermi-technology would be appropriate for your goals. Now let's go over how to begin vermicomposting.

The basic process is to start with about 6 inches (15 cm) of moist bedding. Add composting earthworms at a rate of 1 to 2 pounds per square foot of surface area (0.5–0.9 kg per 0.09 m²). Apply a layer of feedstock 1 to 1½ inches (2.5–3.8 cm) deep. Then wait until the feedstock is consumed before adding more. If you are feeding raw food scraps, be sure to cover them with shredded paper, cardboard, cloth, a tarp, or another suitable material.

I can't adequately emphasize how important it is to follow this basic procedure. However simplistic or obvious it may seem, I've seen countless people miss a step or otherwise mess up during the 25 years I have been practicing, researching, and teaching vermicomposting. So let's go into each step in detail.

Whatever type of vermi-system you use, your worms need organic material, called bedding, to form their habitat. Suitable bedding has four important properties: It retains moisture, provides shelter, is loose in texture, and is porous. Moisture is needed to keep your worms' skin moist. The bedding has to provide shelter from temperature extremes, light, and UV rays. Loose texture allows the worms to move freely. Porosity allows oxygen to infiltrate the bedding, and of course, earthworms need oxygen for life, just as all other animals do.

Starting your system with 6 inches (15 cm) of bedding is important to ensure an adequate amount of space for the worms to live. The bedding should be no deeper than 6 inches because more than that may provide too much mass, and thus the opportunity for the bedding to heat up.

Next come the worms—at least 1 pound per square foot or, if you use the metric system, 0.5 kg worms to 0.09 square meter, of surface area. If you add less than a pound of worms per square foot, your system will take longer to get up and running, and you'll be at greater risk of overfeeding. You can

FIGURE 6.1. I spread food scraps on the surface of my continuous flow-through bin (*top*) and cover them with a layer of shredded paper about 2 inches (5 cm) thick (*bottom*) to prevent the food from attracting insects.

stock your system with up to 2 or even 2½ pounds (0.9–1.1 kg) of worms per square foot, but if you exceed that amount, you're just wasting money and inviting trouble.

Go back and reread the sentence about the stocking rate. The last part of it—"surface area"—is often overlooked despite its importance. Surface area is key because the type of earthworms used for vermicomposting don't burrow, and they move upward through whatever material surrounds them to feed from the surface. Therefore, what matters is the surface area of the bed, not the depth. To determine the total square footage of surface area of a bed, multiply the length of the bed by its width. For example, if your bin is 8 feet long by 5 feet wide (2.4 × 1.5 m), you have 40 square feet (3.7 m²) of surface area. Consequently, you should stock the bed with 40 pounds (18 kg) of worms to get started.

Once you've added worms to the system, you need to provide them with something to eat. The thickness with which you apply feedstock is crucial because of two factors: overfeeding and overheating. You will notice that I mention the dangers of overfeeding repeatedly throughout this book—that's because it is so important! If you put a thick layer of feedstock in the worm bed, the lower part will likely become anaerobic and the worms won't be able to process all of it. A thicker layer also has the potential to heat up, which can kill your worms! I recommend aiming for about a 1-inch (2.5 cm) thickness of feedstock—1½ inches (3.8 cm) maximum—to avoid both overheating and overfeeding.

Do not add another layer of feedstock until the first has been consumed. This falls under that all-important category of overfeeding. If you put fresh food on top of uneaten food, the worms will come to the surface to eat the fresh stuff and leave the older food to rot and become anaerobic. So patiently wait for your worms to finish devouring what you last fed them.

The final point on the list is for those who plan to feed their worms raw food scraps. Raw food will attract flies, gnats, fruit flies—all kinds of annoying, flying insects, and maybe even cockroaches, too. If you keep the raw food covered, though, it will not catch the attention of insects.

Bedding Options and Preparation

Bedding is the organic material your worms will live in. As I mentioned in the previous section, it needs to be stable (meaning not inclined to heat up), have the ability to retain moisture, and remain loose so that oxygen can circulate through it. Bedding should contain very little protein, and you should be cautious of organic nitrogen compounds that quickly decompose—they will release ammonia that is harmful to the worms. If bedding material is not already significantly decomposed, or if it contains a lot of readily degradable carbohydrates, it could heat up and cause the worms to die.

To avoid these pitfalls, stick to using materials that have a proven track record, such as:

- Stable compost that is not high in soluble salts
- Horse manure that has aged for at least three months
- Coconut coir (shredded husk material)
- Shredded paper
- Aged leaf mold
- Shredded brown leaves

Add one of the bedding materials in a 6-inch (15 cm) layer at the bottom of your bin or pile; if you are vermicomposting in an extraordinarily shallow system, the bedding can be as little as 4 inches (10 cm) deep.

Although some worm farmers use peat moss as bedding, I recommend against it. Peat moss is too acidic for optimal worm growth. You have to add lime to achieve the correct pH, and it is difficult to reach the right balance. Why go to the trouble of purchasing peat moss and lime when you can get other types of bedding at no cost? Also, harvesting peat moss in the wild is not an environmentally sustainable practice.

When applied, your bedding should be moist. Soak the bedding in water—ideally at a temperature of 65 to 70°F (18–21°C) to maximize absorption—and remove excess liquid *before* adding it to your worm bin. If you put dry bedding into the bin and then pour water over it, you will not be able to control the moisture content, and excess liquid will run off. This is undesirable for many reasons, and it may also violate composting regulations. For more instructions on monitoring and adjusting bedding moisture, refer to "Moisture Control" in chapter 8 on page 134.

The top of the bed should remain moist, but never soggy, throughout the vermicomposting process. If the top of the bed is not moist enough, use a misting system to apply water. Never *pour* water into your worm bin or bed!

Adding Earthworms to Your Vermi-System

For rapid vermicomposting you should add a minimum of 1 pound (0.5 kg) of earthworms per square foot (0.09 m²) of bed surface area. Some worm farmers report having 2 or 3 pounds (1–1.5 kg) of worms per square foot of surface area (which offers the advantage of more rapid production of vermicast). If it is your goal to have that dense a population of earthworms, I

Worm Breeding

Some worm farmers use shallow trays from the construction or gardening industries to breed more worms, either for their own vermicomposting operation or to sell to others. Breeder worms should be fed a consistent type and quantity of commercial worm food or animal manure. After the mature worms produce an abundance of cocoons, place them in new trays with fresh bedding to begin the process again. The tray containing cocoons should be divided among new trays and covered with fresh bedding and food. The cocoons will hatch, and the baby worms will grow to reproductive maturity and produce new cocoons. Then you repeat the process of harvesting the mature worms, putting them into new bins, and dividing the cocoons into new bins.

Mortar mixing trays and concrete mixing tubs are particularly popular for breeding worms. A large mixing tub has a 21-gallon (80 L) capacity, costs about $13, and is made of heavy-duty resin that is lightweight and flexible. The tub is 36 inches by 24 inches by 8 inches deep (91 × 61 × 20 cm) and weighs just over 4.5 pounds (2 kg). A medium-sized mixing tray is 28 inches by 20 inches by 6 inches (71 × 51 × 15 cm), holds 10 gallons (38 L), weighs under 2½ pounds (1.1 kg), and costs less than $6.

People like using mixing trays for breeding because they can be placed on shelves, and a full tray of worms is light enough to move around by hand. Mixing trays also have an ideal surface-area-to-depth ratio for worm farming, providing the worms with plenty of access to oxygen. Don't stack the bins on top of each other, and pay close attention to the moisture levels because the bedding can dry out quickly without a lid.

suggest that you start with 1 pound per square foot and let the worm population naturally increase.

On average, worms cost $20 to $50 per pound, depending on the quantity you purchase. Because there are so many factors that can cause worms to die or crawl away, it is foolish to buy a large quantity before you have worked out the kinks in your system and learned how to provide a proper environment for your worms. Over the past 25 years, many people have contacted me eager to start a large operation right away with 50 or 100 or more pounds of worms. I tell them to slow down and start with 1 pound of worms in a 14-gallon (53 L) plastic storage bin so they can learn the basics of earthworm husbandry without a large financial investment. (See the resources section for links to information on small-scale vermicomposting.)

When you receive a shipment of worms, check to make sure that all of the worms arrived safely (didn't die) and estimate the total number you were sent. One way to do this is to make a visual estimate of the number of worms: There should be close to 1,000 worms in a pound. For larger quantities, you could try weighing what you receive, but you should still estimate the number of worms, too, because some growers will put the worms in a large quantity of organic material, assuming that there are plenty of worms in there. By visually sorting, you may discover that there are only dozens or hundreds, not thousands.

Some worm growers will either mistakenly or intentionally send you fewer worms than you ordered. If you have been undercut, contact the grower right away and discuss it with them. If you're lucky, they will give you a refund or send more worms. Depending on how the transaction goes, you will have to decide whether or not you wish to order from the same supplier in the future.

Your worms will arrive in a bag, a box, or both. Gently empty the earthworms and bedding onto the top of your worm bed and then step back and watch as the worms quickly move into the bedding to escape the light. Inevitably, not all of the worms will come out of the bag or box, and if you try to remove them, you could accidentally damage their fragile skin. I leave the bag on top of the bedding and allow the remaining worms to crawl out of the dry bag into the moist bedding on their own accord.

Sometimes a shipment of worms will arrive earlier than expected and you will not yet be ready to add them to your vermi-system. In this case unseal the boxes, take out the bags of worms, and empty the worms into temporary trays or bins filled with 3 inches (7.6 cm) of bedding. You can leave the worms in the trays for up to 12 hours before adding them, with the bedding in the trays, to your vermi-system.

When you add worms to your system, bear in mind that they will take time to adjust to their new home. It is highly likely that the worms were raised in bedding that differs from yours, and they have also just endured stressful transport of some sort, whether bouncing around in your car or experiencing a variety of vibrations and temperature swings while being shipped in a box. Shaken up from the ride and placed in an unfamiliar environment, they may initially try to escape the bed. One way to prevent escapes is to leave lights on continuously in the system area for up to two weeks, forcing the photosensitive worms to stay in their new bedding. As time passes, the worms will adjust to the bedding and stay in it on their own accord.

Once your worms have adjusted, they will be ready to feed. Feeding worms is a big topic, which I cover in detail in the next chapter. Before you delve into the feeding process, though, it's important to consider some other vermicomposting basics: setting up a water supply and possibly a source of electric power and establishing good safety practices.

Utility Needs

The two main utilities you will need for your vermicomposting operation are water and electricity. Access to water is an absolute necessity, while electricity may be optional depending on your situation, which is good news for folks who don't have access to it.

Water

Water is essential for keeping your beds moist and providing a hospitable environment to your worms. The chlorine found in most tap water will kill microorganisms, so it is best to collect and use rainwater or well water. If you do use tap water and you're unsure whether your water contains chlorine, pick up a test kit from a swimming pool supply store. If you find that there is chlorine in your water, you have a couple of options. You can install an activated carbon filter that will remove the chlorine, or you can run water into a container and let the chlorine evaporate for 24 or 48 hours prior to use. The chlorine will evaporate faster if you use an aeration device that adds oxygen to the water, such as an aquarium air pump.

Some areas have "hard water," which contains elevated levels of calcium, magnesium, and other minerals. Municipalities in these areas will often chemically soften their water using a brine solution to replace calcium and magnesium ions with sodium ions. The "harder" the water is, the more salt needs to be added, and the higher the final sodium level of the water is. Be careful if you're using chemically softened water to moisten your worm bed—it could make the salt levels too high for your worms. This is a situation where it may become essential to use rainwater or well water rather than tap water.

Some worm growers use automatic misters like those used in commercial greenhouses, which can be programmed to deliver a certain amount of water, in the form of mist, as frequently as six times per day. If you choose this type of system, you do not need to hire someone who has plumbing skills to install it; most people can handle it as a do-it-yourself project.

Other worm growers hand-water their worm beds with a hose and nozzle set for a fine spray, typically once or twice daily, depending on the conditions of the bed. Regardless of whether you hand-water or use automatic misters, you need to monitor the moisture in your beds daily and adjust your watering amount and schedule accordingly. You will find that your beds need more water during one season than another.

Someone recently asked me, "What is the hour-to-liter ratio I should mist my worms during the heat of the summer? And how many times a day should I mist

FIGURE 6.2. At Green Organic in Afghanistan, hand-watering is important to keep the bedding in these concrete-block bins at the right moisture content. *Photograph courtesy of Qasim Bakhshi.*

during the summer heat?" I replied that there is no universal setting and frequency to use for a misting system. The hour-to-gallon or -liter ratio depends on your water pressure and volume and hose diameter and length. The frequency depends on temperature, humidity levels, amount of moisture in the bedding, whether or not your bed has a cover, and other factors. Everyone's situation is different, so take time to tinker with your irrigation system to determine what works best for you.

It is essential to determine your water pressure and volume before designing a system of automated irrigation for your worm bed. To determine volume, you can figure out your water system's gallons-per-minute delivery by timing yourself as you fill a 5-gallon bucket with water. To ascertain your water pressure, you can get a PSI gauge from a hardware store and attach it to your faucet.

Adjust your automated irrigation system after you have determined your particular needs rather than relying on what the instructions recommend. Be sure to adjust the timer according to changing climatic conditions. A control clock can be set for irrigation duration and frequency.

You can connect soil moisture sensors (SMS) to your irrigation system to further refine the process. Simply bury the sensors in the worm bed, and they will provide an estimate of the moisture content. Bypass-type SMS controllers have an adjustable threshold setting that prevents the system from delivering water at scheduled times if the beds are already adequately moist.

Manufacturers make a variety of misting nozzles that will spray water in different angles and patterns according to your needs. For example, certain nozzles spray 360 degrees, some cover 180 degrees, and others can spray in a rectangular shape. Misters that spray in a circular pattern leave too many dry edges on a worm bed, so look for the rectangular spray nozzles.

Some larger vermicast manufacturers use a greenhouse boom irrigation system to uniformly water the beds of their continuous-flow reactors. A boom system has one or more pipes with nozzles that spray water as they move over the bed. The system is either suspended overhead or situated on a mobile cart that moves between the bins. A trailing hose delivers the water, and a cycle timer activates a motor to drive the boom unit.

It's great to have all these technology options, especially if you cannot always be at your vermi-operation every day, but be careful not to rely on them too much. Unanticipated things can and will happen, whether it's the electricity going out, hoses breaking open, timers failing, or sensors not working. If you're away for too long, you might return to your vermi-operation and discover that your bins are too dry or too wet, and then you'll have to try to fix the problems. It's better to stay on top of your operation by checking the bins daily, rather than fix problems after they've had time to become more serious.

Electricity

If you want lighting or hope to run a pump for an irrigation system, you'll need electricity. As mentioned earlier in this chapter, keeping lights turned on can inhibit worms from trying to flee from a new bedding environment. Some worm farmers leave lights on over their worm beds on rainy nights as well, since worms have a tendency to crawl when it rains. In both cases you'll want to ensure that the light bulbs don't cause your worm bed to overheat, so choose energy-efficient bulbs, and don't place them too close to your worm bin. There are no standard rules to follow here; it's a matter of figuring out what lights to use and where to position them in relation to the bed to deter the worms from leaving without heating up the bed. Electricity can also be handy to heat or cool a building that houses the worms; to operate machines and tools; and to control the temperatures of worm beds by running fans, heating pads, or heating coils.

Safety

It is important to protect yourself, your customers, and the vermicomposting industry as a whole. There are a variety of safety issues that would be good for you to consider for your vermi-operation.

Safety from Bites and Stings

For your personal safety, be mindful of the possibility of creatures in or around your worm beds that can sting or bite you, workers, and visitors. I refer to my Worm Barn as "the spider factory" because spiders of all types, including black widows, love the nooks and crannies throughout the building and in the compost bins outside. A number of other creatures haunt the Worm Barn, including hornets. I have been stung many times while simply walking through the barn. Snakes sometimes curl up in a corner of the barn, a pile of leaves, or another type of feedstock.

I found a black rat snake in my 40-square-foot (3.7 m²) worm bin several years ago. As I pulled back the black cover, the snake shimmied across the surface of the bed. Part of my brain was saying, "Of course there is a wiggling creature in the worm bin," and another part was exclaiming, "It's not a worm, it's a snake!" It was surprising, but not a big deal because I like snakes. Nevertheless, some people freak out when they see snakes, so afterward, whenever I gave folks a tour of my Worm Barn, I would rap on the side of the bin before removing the cover. That way, I hoped, the snake would move out of the way before I took the cover off and not scare my guests.

Safety from Pathogens

The vermicast you sell must be safe for plants, people, and pets. Get out of the "It's natural, therefore it's safe" mind-set, and test your products (as explained in chapter 9). Selling vermicast that contains plant diseases, pathogens, or persistent herbicides could cause you to lose customers or your whole business. Could your business survive getting sued by a grower whose crops were killed by your vermicast?

Beyond just harming your business, poor safety practices could very well put a black eye on the whole vermicomposting industry. Several years ago I heard about a worm farmer who sold vermicast containing plant pathogens that killed the crops of several greenhouse growers. Not only was the worm farmer in trouble, but the word was out that "vermicast will kill your plants." Using online forums and Listservs, greenhouse growers and farmers circulated the message, "It is dangerous to use vermicast."

What do I mean by "pathogens"? Pathogens are organisms that can cause disease within another host creature. They are usually bacteria, fungi, viruses, parasites, helminths, protozoans, prions, or other types of microorganisms. The pathogens we hear about most often in the headlines are *Escherichia coli* (*E. coli*) and *Salmonella* spp., which are occasionally responsible for product recalls due to food contamination.

Besides food, other sources of pathogens include yard trimmings, raw manure, and biosolids (sewage sludge). *E. coli*, helminths, and *Giardia* spp. often appear in yard trimmings due to contamination by dog, cat, or bird feces. In addition, yard trimmings can carry plant pathogens.

Raw manure from livestock may contain a variety of pathogens, including *E. coli*, *Salmonella*, *Yersinia*, *Campylobacter*, *Giardia*, *Leptospira*, *Rotavirus*, and *Cryptosporidium*. Some heartier pathogens survive for a lengthy period of time in manure, while others can be easily destroyed by processing techniques or temperature extremes. The majority of bacteria and viruses in biosolids are destroyed when they are processed at wastewater treatment plants, but some can survive.

There are many factors to consider when it comes to preventing contamination by pathogens. For example, how you manage your equipment plays a role, as explained in "Processing Feedstocks" in chapter 7 on page 117.

I'm mentioning these things not to scare you but to encourage you to be mindful of the potential presence of pathogens and to proceed cautiously. Always wash your hands after handling worms or feedstocks, just to be safe!

Respiratory Safety

Aspergillus fumigatus is a fungus that grows on decaying matter, like the bedding and feedstock that you might put into a worm bin. Aspergillus is a common mold found indoors and outdoors, and it's rare for people who breathe in the spores daily to get sick from them. Nonetheless, it can be a big problem, even fatal, for immunocompromised people with low white blood cell counts or folks with lung diseases. When inhaled, its spores can infect the lungs with aspergillosis (often called farmer's lung), which is a form of hypersensitivity pneumonitis.

Last year a pulmonary physician in another state emailed me asking advice regarding a patient. A biopsy confirmed that the patient had hypersensitivity pneumonitis in her lungs, and the doctor was trying to pinpoint the cause. He had deduced that it was caused by inhalation of fungi from her worm bin, and he wanted my opinion about the specific culprit. I shared what I knew about aspergillosis and concurred that it could have been caused by prolonged exposure to her worm bin and feedstocks. His patient's health has since improved through reduced exposure to the worm bin, the use of a mask, and occasional use of prednisone.

As a side note, *Aspergillus fumigatus* also affects bagpipers. The day after the doctor contacted me, there was a news story about a man in Scotland who had just died. He was seriously ill with a lung infection, and the doctors couldn't figure out what was causing it. He took a trip to Australia for three

FIGURE 6.3. If you are immunocompromised, wear a dust mask to prevent inhalation of fungal spores when working with your worm bin. *Photograph courtesy of J. Mark Rice.*

months, and his health improved dramatically. When he returned to Scotland, his health deteriorated again, and he died. Doctors finally determined that his hypersensitivity pneumonitis was caused by his daily bagpipe playing (which he had suspended while he was in Australia). The dark, warm, moist environment inside of wind instruments is ideal for breeding mold and fungi.

Don't be alarmed! This infection is highly uncommon, especially if you are in good health. Still, even healthy people would be wise to wear a dust mask when working with hay and straw and in dusty conditions, as it is possible to develop aspergilliosis. If you have a lung disease or are immunocompromised, it would be good to wear a dust mask when you are working with materials in your vermicomposting operation that could harbor aspergillus spores.

Site Safety

It is important to not only inspect your worm beds daily but also regularly check the rest of your site to ensure safe conditions:

- Check frequently for burrowing animals around buildings, bins, and structures. If you find animals, remove them and repair any damage.

- Remove woody vegetation and noxious weeds growing around structures.
- If you are storing raw feedstocks onsite, be sure that the pile height does not exceed 12 feet (3.7 m) to reduce the risk of fire.
- Examine incoming feedstocks for contamination and hazardous waste.
- Create an operations and maintenance manual.
- Conduct monthly safety site audits and meetings.
- Maintain roadways to, from, and around your facility with stabilizing material, stone, or gravel.
- Inspect floors and walls to see if they need repairs.
- Be prepared for extreme weather events.
- Develop a written action plan for emergencies.
- Do not allow runoff from loading areas and from spills to flow into streams or road ditches.
- Post the names and phone numbers of important contacts. This may include emergency contacts, fire department, sheriff's department, poison control, plumber, electrician, gas and electric company, and feedstock sources.

Personnel Safety

If you have employees, familiarize yourself with the federal Occupational Safety and Health Administration (OSHA). You can learn more about their workplace health and safety guidelines at www.osha.gov. Establish standardized procedures, and provide the proper equipment for protection against dust, sight, noise, and other hazards. Personal protective equipment includes:

- Safety vests
- Hard hats
- Safety glasses
- Hearing protection
- Dust masks
- Gloves
- Steel-toe and -shank boots

Protect your workers from heatstroke and heat exhaustion, as both can lead to illness or death. Establish a heat illness prevention program, and educate workers about the factors leading to heat stress and how to avoid them. Provide cool water close to work areas and encourage employees to drink at least 1 pint (0.5 L) of water per hour.

It is a good idea to hold monthly safety meetings for employees. Choose a monthly safety topic, and cover it in depth so that everyone has a deep understanding. Instruct staff on how to recognize potential problems, which can then be brought up and dealt with in future safety meetings as well.

Equipment Safety

To prevent equipment-related disasters and ensure that your operation runs smoothly, study the safety manuals provided by your equipment manufacturers and create an Equipment Safety Plan, including a preventive maintenance program with procedures and schedules. Determine potential work area hazards and identify them with bright-colored paint, tape, or warning signs. Mark off "throw zones" near grinders or other machinery that could shoot out fast-moving airborne objects. Label hazard areas, too, such as pinching, crushing, and limb-loss hazards where a person's body could be caught between the moving parts of chains, belts, cables, or other machinery. Identify sources of possible electrocution, and implement a lockout-tagout (LOTO) system to prevent equipment from being started up while someone is carrying out maintenance or repairs on it.

If fuel is delivered to your site (to power vehicles and machinery), develop a fuel delivery and containment plan (see the resources section at the end of this book). Determine whether you are required to provide secondary containment for bulk storage containers, and create a schedule for inspecting and maintaining fuel containment systems.

Fire Prevention and Safety

If your vermicomposting operation stores large quantities of raw organic materials, there is the possibility of a fire starting—it's much more common than you may realize. If a pile catches fire, it may take *weeks or months* to put it out and cost hundreds of thousands of dollars. For example, I recall a fire at one composting site that took three weeks to put out and clean up, at a cost of over $375,000. Ignition sources include cigarettes, sparks from equipment or vehicles, heat from grinders, electrical fires, lightning strikes, wildfires, arson, or spontaneous combustion from raw feedstock piles.

Spontaneous combustion occurs when heat is generated within a pile at a faster rate than it is lost to the environment. The temperature continues to rise until it reaches a range of 160 to 200°F (71–93°C), setting off a chain of heat-generating reactions that ends with spontaneous combustion. The conditions that allow spontaneous combustion often develop within huge piles of mostly dry organic material that contains 20 to 40 percent moisture (with periodic wetter areas) and that has been ignored for several weeks. It can happen not only to raw feedstocks, but also to compost piles that are curing, or even to finished compost piles.

Feedstocks that have self-combusted at composting facilities include wood chips, mulch, sawdust, bark, pallets, leaves, branches, and grass. It is wise to initiate a meeting with your local fire department as you plan your vermi-operation. Describe the types of raw organic materials you will have onsite, and if you will be pre-composting, tell them so. Come to agreement on how you and they will manage an organic materials fire if it begins. Post a Fire Response Plan onsite next to appropriate firefighting gear such as extinguishers, water, and hoses. Incorporate wind- and firebreaks, and maintain access roads for emergency vehicles.

Prevention is the best way to avoid a dangerous and expensive fire. Remove dust, and store flammable materials away from areas and equipment that might produce sparks. Install lightning protection apparatuses. Avoid pile depths higher than 12 feet (3.7 m), and ensure sufficient ventilation throughout the piles. Regularly check the temperatures of your piles in several places, and look out for steam vents (openings in the pile surface with steam emerging from them) in particularly deep piles. Determine the hottest spot in your piles and manage them before a fire breaks out. You can do this by carefully breaking down particularly hot piles into smaller piles.

Following Safe Work Practices

I've covered possible safety hazards that you should be aware of at your vermicomposting operation. You would be prudent to adopt strict workplace rules for anyone working at your vermicomposting facility. There are numerous ways that workers can either get hurt or cause harm to your operation. Here are some rules you should consider adopting:

No smoking. You will likely be working with organic materials that can easily combust, so adopting a no-smoking rule will reduce the chance of a fire.

No cell phone use while on the clock. Cell phones can be a distraction and lead to carelessness on the job, with consequences including injury, death, and ruined machinery.

Wear safety glasses. Certain tasks could result in injury to eyes, such as chopping or grinding organic materials, as well as drilling, hammering, sawing, sanding, soldering, or welding. Moving organic materials can also result in dust or particles flying into eyes. Require that workers wear eye protection whenever they perform one of these tasks.

Wear gloves. Work gloves should be worn to protect hands from cuts, and sanitation gloves should be used when handling pathogenic or toxic materials.

NO DRUGS OR ALCOHOL. Requiring workers to show up sober for their job will reduce the chance of injury and increase productivity. Make sure that workers inform you if they are taking medicine that warns against driving or using machinery.

PRACTICE FIRE SAFETY. Ensure that every worker knows where firefighting equipment is located and how to use it.

LIFT CORRECTLY. Instruct workers on how to lift with their legs, not their backs. Require them to ask for assistance if the load is too heavy to handle alone.

USE SAFETY DEVICES. Do not disconnect, remove, or destroy safety mechanisms on equipment with moving parts.

TURN OFF MACHINERY when unattended, or if you need to lubricate or fuel it.

INSPECT EQUIPMENT REGULARLY. Set up a schedule to inspect items such as ladders, tools, or machinery, and remove defective items from service.

PRACTICE ELECTRICAL SAFETY. Check the electrical grounding system daily. Always use ground fault circuit interrupters, especially when tools are being used with temporary power supplies or in wet conditions.

PRACTICE EMERGENCY PREPAREDNESS. Post your emergency procedures, and be sure all workers have read them. Make sure every worker knows the locations of first-aid kits and how to get urgent assistance.

KEEP EVERYTHING IN ITS PLACE. Assign specific places for tools, ladders, hoses, and machinery, and always return them there after use so they do not become tripping hazards.

FOLLOW SAFE EQUIPMENT PRACTICES. No hitching rides on equipment such as skid loaders or tractors—if there is not a seat for you, do not get on it.

CHAPTER 7

Earthworm Feeds and Feeding Methods

W orms will eat compost, leaves, manure, food scraps, shredded paper or cardboard, coffee grounds, spent mushroom waste, brewery residuals, agricultural crop residues, grains, food processing waste, and more. But bear in mind that although I say that worms "eat" leaves and other items, what they actually consume are microorganisms and tiny particles of organic materials. Microorganisms are the first to dine on much of the feedstock material added to a worm bin or bed. Remember, too, that just because your worms can consume something, that doesn't mean they should. Don't think of them as living garbage disposals that you can feed anything. Worms need a balanced diet for their health and the quality of their castings.

Feedstock Options

As you plan your vermicomposting operation, consider the feedstock options available to you within a reasonable distance. Your plan for selling or using the end product should help you decide which feedstocks to use. As noted earlier in this book, if you want to sell vermicast to end users who require consistent quality, you need to be able to obtain a long-term supply of appropriate feedstocks to continue producing uniform vermicast. You will risk losing customers if you run out of one type of feedstock and have to switch to another that results in lower-quality vermicast.

Determine whether acquiring the feedstocks will be an income source or an expense. In other words, will someone pay you to take their waste, will they give it to you for free, or will you have to pay for it? My inclination is to first look for

FIGURE 7.1. Food waste can be an excellent feedstock for vermicomposting. Possible sources include restaurants and institutions such as nursing homes and colleges. *Photograph courtesy of Zachary Brooks.*

a good-quality feedstock that I can obtain at no cost or—better yet—that someone will pay me to take. A desirable feedstock may be considered a nuisance by someone else. For example, are there nearby farms that consider their animal manure as waste and will pay to have it removed? Will they deliver it to you to get it off their hands? Consider nearby industrial operations, too, such as paper recycling plants, lumber mills, breweries, mushroom producers, food processors, or food distribution operations. Check out other food waste generation sites, including prisons, hospitals, nursing homes, military bases, daycare centers, restaurants, schools, colleges, and large office buildings. Perhaps you're even lucky enough to already own a farm that generates adequate livestock manure, bedding, or crop residue for your vermicomposting operation.

Keep in mind that to produce a high-quality, consistent vermicast product, you will need reliable sources that generate copious quantities of feedstocks that have the correct balance of carbon and nitrogen (25:1 to 35:1 is ideal).

Livestock manures, including those from cows, swine, sheep, goats, horses, llamas, alpacas, and rabbits, are all excellent choices for vermicomposting feedstock. The notable exception is poultry manure, which contains twice as much ammonia as dairy cow manure. Chickens excrete a mixture of urine and manure that contains high levels of nitrogen and mineral salts. If chicken manure is fed straight to worms without pre-composting or dilution with other feedstocks, it could kill them.

Manure is dense and high in nitrogen. Mixing in silage, hay, straw, wood chips, or rice hulls adds carbon to produce a balanced carbon-to-nitrogen ratio. These materials will also add structure to the manure and increase airflow and porosity.

Human feces, euphemistically called humanure, can be consumed by earthworms. In fact, it is already being used as a feedstock in numerous countries worldwide. In Australia vermicomposting of humanure and biosolids is common at homes, businesses, and wastewater treatment plants (WTP). In the United States, some people have begun adding worms to their composting toilets, but vermicomposting has yet to become popular at wastewater treatment operations.

Research on the vermicomposting of sewage sludge, or biosolids, in the United States began in the late 1970s at the State University of New York (SUNY) in Syracuse and has been continued by various scientists in the decades since. A famous study was conducted almost 20 years ago at a wastewater treatment plant in Florida, demonstrating that vermicomposting reduces pathogens to safe levels. However, when the results were presented to the US Environmental Protection Agency (USEPA) to seek approval for Class A designation, they were rejected for not meeting the agency's research protocols. Other studies have since been published in scientific journals further demonstrating that pathogen levels are reduced via vermicomposting, but more research needs to be conducted to reach a consensus in the field.

Cow Manure

Because beef cattle and dairy cows are raised for different purposes, the husbandry practices for each differ, and so does the resulting manure. I am not aware of any worm farms in the United States that are using manure from beef cattle, and I've been told that it is harder to vermicompost. Dairy manure is the preferred feedstock of the majority of large vermicomposting operations in the United States. It is nitrogen-rich, it decomposes quickly, and earthworms will readily consume it whether it is raw or pre-composted.

Horse Manure

Worm farmers use horse manure as either a bedding or a feedstock. When used as bedding, it should be aged for at least two or three months so that it will not heat up. Horse manure typically has a great deal of high-carbon material already mixed in (because it is used as bedding for the horses), and the mixed material is relatively dry and results in a higher carbon-to-nitrogen ratio than manure alone.

Swine Manure

In the United States not many worm farmers have used swine manure as feedstock. The only one I'm aware of lived in North Carolina and produced a high-quality vermicast (he is now retired). Swine manure is usually flushed out of hog houses with water. It's stinky, high in nitrogen, and very wet. Due to the pigs' diet, their manure may also be relatively high in salts and ammonia. Running swine manure through a solids separator will squeeze out much of the liquid and make it suitable for feeding to earthworms either raw or as pre-compost. Solids separators typically cost tens of thousands of dollars, and although they can be located outdoors in mild climates, they need to be sheltered or kept indoors in cold environments.

Food Processing Wastes

Using food processing wastes as a feedstock is tricky, so proceed with caution. They come in a wide variety of forms, ranging from wet, nitrogen-rich wastes to relatively dry and carbonaceous filter press cakes. Have food waste tested at a laboratory to determine the carbon-to-nitrogen ratio and moisture content, and then decide if it will be appropriate for your end-use plans. (See chapter 9, page 163 for information about materials testing.)

Vegetable and Fruit Residuals

An increasing number of worm farmers are vermicomposting discarded fruits and vegetables. These wastes vary widely in content. Some degrade quickly, and others have tough skins that are resistant to decomposition. They have a low to moderate carbon-to-nitrogen ratio and a medium to high moisture content. There may be sticky labels, plastic bags, and other contaminants mixed in.

A note about citrus: It's usually okay to add some to a large-scale vermi-system. I won't add it to my 14-gallon (53 L) household-sized worm bin, but I haven't had a problem putting it in my 40-square-foot (3.7 m²) continuous-flow reactor.

Paper

Printer paper and newspaper are often used for vermicomposting, both as bedding and as feedstock. They are dry and carbonaceous, so they need to be mixed with a moist, nitrogenous material to balance out the C:N ratio. Be aware that paper has poor structural integrity and can become mushy, which inhibits airflow.

Worms Eat My Dust

Paper Shavings, a paper recycling business in Greeley, Colorado, was grinding paper to make animal bedding. The company generated a tractor-trailer load of paper dust every four to six weeks. They decided to vermicompost some of the dust onsite and send the rest of the paper scraps to a composting facility. They devoted more than half the space in their 25,000-square-foot (2,325 m²) warehouse to setting up a dozen worm bins that were 9 feet long by 5 feet wide (2.7 × 1.5 m), and made plans to expand to 40 bins. They hand-built the bins by repurposing wooden pallets. To provide an optimal environment for the worms, they heated the building to 75°F (24°C). They developed a recipe of 30 percent paper dust and 70 percent green waste compost blended in a hydraulic mixer they built and attached to a forklift. They harvested 150 to 200 pounds (68–91 kg) of vermicast every 10 days.

People often ask me whether paper printed with ink is safe for vermicomposting. Inks used to contain heavy metals, and toxic chemicals were used in the papermaking process, but once people realized the environmental impact, federal regulations were established to limit toxicity. Manufacturers now make most ink from soybeans or vegetables, and paper is often bleached using oxygen instead of poisonous chemicals, so most office paper—whether printed with ink or not—should be fine to use in worm bins. I do discourage people from adding certain types of paper to worm bins, however. Glossy magazine paper, for instance, can contain clay with heavy metals, as can greeting cards decorated with glitter.

Paper Mill Sludge

Paper mill sludge is produced during the final stages of the papermaking process. It contains cellulose fiber from recycled paper or wood, has a moisture content of approximately 45 to 80 percent, and possesses a medium to high carbon-to-nitrogen ratio. Paper mill sludge should be mixed with a dry nitrogen source to provide structure.

Sewage Sludge and Septage

High in nitrogen and moisture content, sewage wastes need to be mixed with carbonaceous material to provide structure for aeration and to balance the carbon-to-nitrogen ratio. The carbon amendment may need to be two to four times the volume of the sludge or septage. Sewage wastes can contain heavy metals and human pathogens, so you should proceed with caution if you choose to use them. In the United States there are regulations regarding the management and processing of these materials. If the sewage has been processed in a wastewater treatment plant, it will contain chemicals that can kill worms, so you should always hot-compost the material to break down the chemicals before feeding it to your worms.

Cardboard

You can use cardboard for vermicomposting, but it's important that you use the right type. Avoid the thick, paper-based material called paperboard, which is used to make cereal boxes and other consumer goods packaging. You are likely familiar with paperboard because you have opened the flaps and flattened these boxes for recycling.

For vermicomposting, use corrugated cardboard, also called containerboard. Corrugated cardboard has three layers of paper—two flat sheets and a pleated liner glued in between them—which gives it strength so it can be used to create shipping containers. This type of cardboard is high in carbon content and the glue provides nitrogen, making it more nutrient-rich than paperboard.

Developing a Feedstock Recipe

Now that we've discussed feedstock options, it's time to figure out how to combine organic materials into an appropriate feedstock for your earthworm population. This is called developing a feedstock recipe. Sometimes a single type of material, such as paper, doesn't provide the ideal types of nutrients in the proper amounts for worms, so you may need to mix different kinds of materials together. One important benchmark for feedstock quality is its ratio of carbon to nitrogen, so let's look at that in more detail first.

Some of you may be wondering what I mean by carbon-to-nitrogen ratio (C:N). Remember in high school chemistry when you were introduced to the periodic table of elements? It is a list of 118 chemical elements, each with one- or two-letter symbols, plus an atomic number (the number of protons in the element). C:N is the ratio of total mass of elemental carbon (C) to total mass of elemental nitrogen (N) in a substance. A feedstock's C:N ratio indicates how much more carbon than nitrogen exists in the organic material. Nitrogen is always expressed as "1" (N = 1) in the ratio. Therefore, to represent how much more carbon there is in relation to nitrogen, the value will be x:1.

C:N Ratios of Common Vermicomposting Materials

The following are carbon-to-nitrogen ratios (calculated on a dry-weight basis) for a variety of organic materials that you may consider using in your worm beds. The materials are listed from highest to lowest carbon content, and conversely from lowest to highest nitrogen content.

Wood chips 200:1–700:1	Cornstalks 60:1–73:1	Sheep manure 16:1
Newspaper 560:1	Paper mill sludge 54:1	Food scraps 15:1–25:1
Cardboard 200:1–500:1	Apple pomace 48:1	Horse, llama, donkey,
Mixed paper 150:1–200:1	Cranberry filter cake 31:1	alpaca manure 15:1–25:1
Wheat straw 140:1–150:1	Leaves 30:1–80:1	Alfalfa 15:1
Rice hulls 121:1	Peat moss 30:1–65:1	Apple filter cake 13:1
Sawdust 100:1–230:1	Garden debris 20:1–60:1	Cottonseed meal 7:1
Triticale, rye, or oat straw	Cow manure 20:1	Swine manure 5:1–12:1
70:1–90:1	Coffee grounds 20:1	Wastewater sludge 5:1–10.1

The ideal starting range for feedstock carbon-to-nitrogen ratio is between 25:1 and 35:1. To reach this ratio, you may need to combine a material with a high C:N ratio with one that has a low C:N ratio.

As you can see from the "C:N Ratios of Common Vermicomposting Materials" sidebar, most materials do not have a precise C:N ratio—there are a range of possibilities. Wood chips have the largest range, 200:1 to 700:1. This is because different tree species differ in their nutrient makeup. To determine the specific carbon-to-nitrogen ratio for locally available wood chips, send a sample of the material to a laboratory. They will analyze the carbon-to-nitrogen ratio, nutrients, moisture content, bulk density, and other characteristics. Labs that test vermicast will usually also analyze raw organic materials that are intended to be used as feedstocks. (For more information about lab testing, see chapter 9, page 163.)

Although it's not as accurate as a laboratory analysis, you can use an online compost calculator to roughly determine the C:N ratio, moisture content, and bulk density of different combinations of feedstocks. See the resources section at the back of this book for listings of some online compost calculators.

Keep in mind that carbon-to-nitrogen ratio is just one factor affecting the degradability of organic materials. Physical factors that affect the rate of decomposition include particle size, structure/rigidity, porosity, and bulk density. 1

discuss particle size in depth later in this chapter, but for now keep in mind that smaller particles have greater surface area, making nutrients more readily available to microorganisms. If there is not enough porosity in the mixture, it can lead to anaerobic conditions.

Manure, biosolids, and food waste are all dense and high in nitrogen. To increase their porosity and allow airflow, it is important to thoroughly mix in wood chips or another carbonaceous material. This added material should have structural rigidity that stands up to the high levels of moisture in most feedstocks. Let me give you an example. Food waste contains a high percentage of water, which prevents the flow of air, so it can become stinky and slow the decomposition process. If you mix in shredded newspaper, the added carbon will balance out the C:N ratio, but the paper will turn mushy, which will close up the pore spaces and reduce airflow. Wood chips, on the other hand, have a rigid structure that prevents the feedstock from matting and clumping together, while still providing an increase in carbon content.

When it comes to bulk density, aim for approximately 800 to 1,000 pounds per cubic yard (365–455 kg/0.8 m³). Materials with a lower bulk density provide more porosity for airflow.

Wood chips and other high-carbon materials will take a long time to decompose, but you shouldn't worry about it—they can be screened out at harvest time and mixed into subsequent feedings. If you are using wood chips in a

FIGURE 7.2. Under this tree at Big Red Worms in Nebraska are four large compost bins where a mix of manure, wood chips, and food wastes are pre-composted. Compost is cured in piles next to the bins. *Photograph courtesy of Kate Picard.*

continuous flow-through reactor, make sure the particles are small enough to be harvested through the bottom grate.

One example of a feedstock recipe from a commercial operation is Big Red Worms in Nebraska. Jeremiah Picard mixes food waste with horse manure bedding and wood chips. He has an 80-by-30-foot (24 × 9 m) concrete pad with four aerated static pile compost bins that each hold 20 cubic yards (15 m³) of material. He composts the mixture for about 40 days, turning the material in the bin halfway through. He uses some of the pre-compost as his feedstock for his worms; the rest of the compost is cured and then sold.

After you develop what appears to be a suitable feedstock recipe, it's important to test it on your earthworms, and I describe how to do so later in this chapter.

Processing Feedstocks

It is important to develop a plan for collecting or accepting deliveries of raw organic materials to be used as feedstocks. You need a designated spot for deliveries, and you should be prepared to process the material in a timely manner. For example, feedstocks that are odorous or full of liquid need to be mixed with dry, carbonaceous bulking amendments quickly before the odors and fluids disperse. Do not accept odiferous feedstocks at the end of the day, because you will not have adequate time to process them.

Depending on the scope of your operation, you may need equipment for loading, transporting, unloading, processing, grinding, mixing, pre-composting, screening, bagging, or storing feedstocks. I recommend calling or visiting vermi-operations similar to what you envision to inquire what type of equipment they are using. Ask lots of questions, such as how long they have been using each piece of equipment and what they think of it. Did they try other equipment or brands before? What equipment would they purchase next time? What do they not recommend using?

FIGURE 7.3. When food wastes are filled with liquid, it's important to quickly mix them with dry material to prevent runoff and development of unpleasant odors.

FIGURE 7.4. The Worm Farm in California uses this front-end loader to move a variety of materials.

Various types of equipment are used for vermicomposting. Two popular types of machines are skid steers and tractor-mounted loaders. A skid steer, also known as a skid loader, is a four-wheel vehicle that can lift, push, dig, or pull materials. You sit inside a rigid frame and operate lift arms that cradle a bucket loader or other attachments. Skid steers are more maneuverable, compact, and lightweight than other types of loaders. A tractor-mounted loader costs more than a skid steer. To decide what would work best in your situation, do some research to learn what capabilities each one has.

A tractor-mounted loader goes by many names, including front-end loader, bucket loader, skip loader, payloader, wheel loader, and front loader. It is a heavy piece of machinery that can move or load materials into a truck or feed-hopper or onto a conveyor belt. It has two arms that extend from the tractor body, and you can attach a variety of devices, depending on your needs: a bucket to scoop up loose materials and load them, forks to lift pallets or shipping containers, and a bale grappler to move hay or straw bales.

It's wise to use different pieces of equipment to prevent "clean" finished materials from getting contaminated by "dirty" raw materials. Designate as "dirty equipment" a loader that only handles raw feedstocks. A different "clean equipment" loader can be used for managing material that has been pre-composted or vermicomposted. If you can't afford to buy separate pieces of equipment, you can sanitize the loader bucket every time you handle raw feedstocks.

Depending on what feedstocks you are using, you may also need a piece of equipment to shred, chop, grind, or chip the organic material. Shredders are often used for cardboard, newspaper, and yard wastes; grinders are helpful for food wastes; and some folks use a straw-bale chopper for straw bales, hay bales, or cardboard. The type of equipment you need will depend on the size of your operation and whether you need to reduce the particle sizes of your feedstocks. Renting shredding or grinding equipment can be a good choice if you need it only periodically.

Testing New Feedstocks

Many commercial vermicomposters have experienced the misfortune of applying a new feedstock only to discover that it kills their worms or drives them away. Even folks who have been successfully vermicomposting for years have contacted me, stunned, to tell me their worms have abruptly died! One of the first things I ask is whether they've changed feedstocks recently.

One group of worm farmers I spoke with experienced this misfortune when the farm from which they'd been getting dairy manure closed, and they were forced to switch sources. They found it hard to believe when I told them the new manure was what killed their worms, protesting that it was dairy manure, just like their worms had been consuming for years. I explained that because the manure came from another farm, there were possible differences: The new farm might use different feed or bedding for their cows, or manage the manure differently. Each of these practices will cause the manure to be different from the previous source.

To test a feedstock's suitability for your earthworms, place the organic material in a small container along with about a dozen worms and observe their behavior over the next 12 to 24 hours. If the worms consume the material, it's appropriate to use as feedstock. If the worms crawl away or die, the feedstock is not suitable. You can also try pre-composting to see if that fixes a problematic feedstock. Continue to experiment with the substance until you are certain that it can safely be fed to your worms.

Worm Feed Characteristics

It's crucial to understand the characteristics of feedstocks that can have an impact on your worms. Those characteristics are particle size, homogeneity, ability to heat up, and the presence of persistent herbicide residues, pathogens, seeds, or deworming medications.

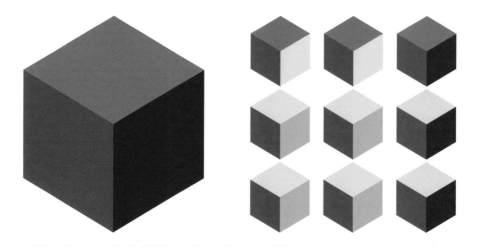

FIGURE 7.5. Chopping or shredding feedstocks creates more surface area overall. The yellow areas on the small particles show new surface areas that become available to microorganisms when the large particle is split into smaller pieces.

Particle Size

As discussed in chapter 3, decomposition takes place on the surface of particles when microorganisms secrete enzymes. Smaller particles have more surface area, which makes nutrients and energy more readily available to microorganisms. As a result, tiny pieces break down more quickly than larger chunks. Figure 7.5 demonstrates how surface area increases when particles are chopped up.

You can reduce feedstock particle size by chopping, grinding, or shredding, or you can run feedstock through a blender, a detached sink garbage disposal, or a pulper.

Don't get carried away with reducing particle size, though. If all the particles in a feedstock are very small, they pack together, reducing the pore size of the organic material and preventing air from being able to circulate. It's important to leave some larger, rigid particles to let in air.

Homogeneity

Worms are picky eaters. Worm growers, home vermicomposters, and science fair students all know this well. To demonstrate worms' food preferences in science fair projects, students have placed four types of food waste in separate mesh bags in different areas of a worm bin and then counted the number of worms in each bag. They usually find that the largest number of worms are in the bag that contains watermelon.

Several years ago a university colleague of mine who worked with straw decided to do an experiment with vermicomposting. He wanted to feed *Eisenia fetida* a mixture of dairy manure and straw. I knew what would happen, and sure enough—just as I predicted—the worms consumed all of the manure and left the straw.

It's easy to slide into anthropomorphic conjecture regarding worms' feeding practices and assume that they consciously have food preferences. In reality, I think it mostly comes down to microorganisms, structure and composition of feedstock cell walls, and feedstock particle size. Individual organic materials will break down more quickly in the worm bed if they consist of simple sugars, have inherently high microbial activity, or are tiny particles that can be sucked in by worms directly.

One of the advantages of pre-composting is that it turns heterogeneous organic materials (with different forms of carbon, nitrogen, and minerals) into a homogeneous substance that your worm workforce will more quickly consume. For example, if you pick up food residuals from restaurants or grocery stores and apply them to your worm bed without pre-composting, your worms will beeline for the melons and stay away from the onions. If you pre-compost the food waste, on the other hand, all of the different scraps will dissolve into a homogeneous soil-like substance that the worms will evenly consume.

Ability to Heat Up

As discussed in chapter 3, heat is generated when microorganisms consume organic matter. If your vermi-system heats up too much, your worms are bound to overheat and flee the bed, so you need to be very careful. Applying feedstock in a thin (about 1 inch/2.5 cm) layer will reduce the risk of overheating, and so will pre-composting. During the winter, you can increase the heat in your worm beds by applying feedstock in a slightly thicker layer.

Another technique to try is to leave certain zones of your worm bed free of feed. That way, if the feedstock heats up (or is too high in ammonia or salts), the worms can congregate in the "safe zones" where there is no harmful food. For example, in my continuous flow-through bin, I designated a feed-free zone that is a couple of inches wide all the way around the perimeter of the bin and another that ran down the center for the full length of the bin.

Residues of Persistent Herbicides

Herbicide residues are a very serious problem—they have crippled large-scale composting operations—so please learn as much as you can and proceed cautiously. As you know, herbicides are chemical substances used to kill weeds

FIGURE 7.6. Pre-composting in aerated static piles like these at Big Red Worms in Nebraska can help to break down toxic materials that may be present in raw feedstocks.

and unwanted plants. What you might not know is that some herbicides can persist on foliage and in the soil for months or years. These are called persistent herbicides, and there are four of them: clopyralid, aminopyralid, aminocyclopyrachlor, and picloram. Some of the products that contain these herbicides include Grazon P+D, ForeFront, Curtail, GrazonNext, Milestone, Redeem R&P, Surmount, Confront, Lontrel, Stinger, Clopyr AG, and Millennium Ultra. Persistent herbicides are used in horse pastures, hayfields, grain crops, golf courses, rights-of-way, and lawns to kill certain broadleaf weeds such as dandelions, clover, and thistle. Horses and livestock can consume these herbicides without being harmed, but they remain active in the animals' urine and manure even after passing through their bodies.

Persistent herbicides must be exposed to sunlight, soil microorganisms, heat, and moisture in order to be broken down into less harmful substances. Sunlight is the most important ingredient, and also the one most lacking during the composting process. This is a problem because these herbicides can remain active in hay, straw, grass clippings, manure, and compost, and they could survive the journey through worms' bodies, untouched by sunlight, and end up in their castings.

Symptoms of Herbicide Damage

The specific ways in which persistent herbicides damage crops include:

Cupping or twisting of leaves
Stunted growth
Main growth tip stops growing and
    lateral buds begin to grow

Reduced yields
Failure of secondary leaves to grow after
    the seed leaves emerge
In legumes, compound leaves stay single

When mulches, manures, composts, or vermicasts contaminated with persistent herbicides are applied to gardens and fields, they can have devastating effects on some broadleaf crops, vegetables, fruits, and flowers. They are particularly damaging to beans, carrots, clover, cotton, cucumber, dahlias, daisies, eggplant, grapes, legumes, lettuce, marigolds, mushrooms, peas, peppers, petunias, potatoes, pumpkins, roses, spinach, squash, strawberries, sugar beets, sunflowers, tobacco, tomatoes, and watermelon.

How long do these herbicides persist? They can be deactivated in 30 days if they receive adequate sunlight and the conditions are just right, but active herbicide has been detected in hay after *three years* of storage in dry, dark barns. Clopyralid has been known to impact plants at 3 parts per billion in soil, even after 18 months in a compost pile. See the resources section at the back of this book for more information about persistent herbicides and the damage they can cause.

Keep this information in mind, and thoroughly investigate potential sources of feedstocks for your worms. If you sell vermicompost that harms crops, it will not only hurt your business but will harm the reputation of the whole vermicomposting industry. You do not want farmers, nurseries, and other growers saying, "Don't buy vermicast—it will kill your plants!"

Contamination by Pathogens

I cringe when I see vermicast and compost advertised as "naturally safe." Too many people think these substances are safe because they employ a natural process of decomposition. The truth is that pathogens may be lurking in any batch of vermicast—or compost that hasn't heated sufficiently—and these pathogens can harm people, plants, and pets. I recommend you reread "Safety from Pathogens" in chapter 6 to be sure you understand the harm pathogens can cause. Pre-composting, which I cover later in this chapter, is one technique that can help reduce the risk of contamination by pathogens.

Presence of Seeds

Food wastes may include seeds, which will take a very long time to decompose and can often sprout in the warm, moist environment of your worm bed. Although this doesn't cause harm to your bed, you will not want seeds in your final vermicast product, because they will sprout where they are unwanted. If possible, remove seeds from food waste before putting it into your worm bed. Seeds can be present in animal manures and yard waste, too, and it would be difficult if not impossible to remove them. The best way to ensure that seeds present in feedstocks cannot sprout is to kill them via thermophilic (high temperature) pre-composting.

Residue of Deworming Medications

Worm parasite control programs are used for a variety of farm animals to eliminate tapeworms, roundworms, lungworms, and other parasites. A group of antiparasitic drugs called anthelmintics (or antihelminthics)—which includes ivermectin, fenbendazole, and pyrantel—is among the most popular for this purpose. Active ingredients in dewormers from other classes include levamisole, dichlorvos, and piperazine. These medications can also kill earthworms.

These deworming medications can be deactivated if animal manure is composted or exposed to sunlight, but if manure is left inside structures or within a large pile, it will take longer for the drugs to be neutralized. To speed up the elimination of dewormers, either compost the manure or spread it outdoors where it can be exposed to sunlight and rain.

Pre-Composting Feedstocks

I highly recommend pre-composting feedstocks for worms. Pre-composting has several benefits, including:

- Killing pathogens and reducing vector attraction
- Destroying weed (and other) seeds
- Reducing weight and volume
- Making the worm feed homogeneous
- Reducing the heat in feedstocks
- Lowering ammonia levels
- Decomposing potential toxins, including dewormers
- Moderating pH levels

- Lowering oxygen demand
- Reducing odor potential

How does pre-composting differ from composting? The two processes begin the same, with the feedstocks heating up to (ideally) 140 to 150°F (60–66°C) so they can achieve the benefits listed. After three or four weeks, much of the material will have decomposed and the temperatures in the pile will begin to drop. To produce mature compost that won't "burn" plants, you would allow the pile to cure for several weeks or months. However, if your intention is to feed the compost to worms, you remove it from the pile after the temperatures drop and you feed thin layers of it to the worms. This is called pre-compost because it is not fully cured or matured. Its value for feeding worms at this point of the process is that it contains an abundance of microorganisms and nutrients that would not be available at such high levels in cured compost.

In a compost pile the activity of microorganisms (bacteria, fungi, rotifers, protozoans, and actinomycetes) consuming organic materials generates heat. During composting, the pile can reach temperatures upward of 170°F (77°C), although it's best to keep the temperature below 160°F (71°C). When managed correctly, composting is effective at killing dangerous pathogens in yard trimmings, animal manures, food residuals, biosolids, and other organic materials. The abundance of beneficial microorganisms in compost also inhibits pathogens from recolonizing the pile once they have been killed off. Temperatures at or above 131°F (55°C) can kill pathogens, and temperatures around 140°F (60°C) will destroy most seeds. Federal and state composting regulations require permitted composting operations to sustain specific temperatures for a certain length of time to kill pathogens.

There are many sources of information to help you learn how to compost. I have authored brief publications that describe how to compost on a large scale and at a smaller level. The US Composting Council has several composting resources, and Cornell University has additional information. When searching for resources online, seek sites that end in .edu or .gov for unbiased information.

Three of the most common methods of pre-composting are in a vessel or an aerated static pile (ASP) or in windrows (these methods are described below). For in-vessel and aerated static pile composting, the rule is to maintain temperatures at 131°F (55°C) or higher for three days. For windrow composting, the rule is to maintain temperatures at 131°F or higher for at least 15 days, and to turn the compost at least five times during that period.

Over the course of three or four weeks, microorganisms consume the readily available energy contained in the organic materials (causing the pile to heat up), and then the compost pile begins to cool to a level slightly higher than ambient temperature. If you are making conventional compost as an end product, you should allow the pile to cure for the next six weeks to six months or longer to eliminate inhibitors to seed germination and crop growth. However, if you're pre-composting to provide feed for your worms, now is the time (when the temperatures have dropped to a stable level) to remove the compost from your pile or bin and let it cool down so you can begin feeding it to your worms. Pre-composted material contains more energy and nutrients for your worms than fully cured compost.

The aerated static pile method can bring compost to a temperature of 140°F (60°C) within a few days. Instead of requiring that you manually turn

FIGURE 7.7. Perforated pipes are part of an aerated static pile at NC State University's Compost Learning Lab in Raleigh, North Carolina. All of the pipes will be covered with at least 12 inches (30 cm) of wood chips, and the primary feedstock (food waste, horse or cow manure) will be layered on top and capped with wood chips.

the compost pile to introduce oxygen and maintain aerobic conditions, an ASP system uses an electric blower to create even airflow throughout the pile. ASP systems can be built by running perforated PVC pipes strategically placed underneath a compost pile and connecting the blower to the pipe system, or by blowing air up through the floorboards beneath a bin.

A 1½ cubic-yard (1.1 m³) ASP bin will hold enough feedstock to feed a 40-foot-long (12 m) worm bin, which provides a good point of reference for size, but keep in mind that you may need to use a skid steer with a bucket loader to manage the organic materials, depending on the scale of your operation. When constructing an ASP bin, consider the type of equipment you plan to use, and make sure that the dimension of the bin will accommodate the machinery.

The most important aspect of operating an ASP system is determining how long to blow air into the bin or pile. You will need to tinker with the system to figure out the best timing for the blower, which depends on your choice of feedstocks and several other factors. I know one worm grower using dairy manure who runs the fan in his 1½ cubic-yard aerated static pile bin 3 minutes on and 12 minutes off throughout the day.

Some worm farmers manually turn their aerated static piles after one week, because turning helps to break down the particles. If a compost bin approaches 160°F (71°C), you should turn on the blower to cool it down to 145°F (63°C).

Many ASP systems have a plenum—a chamber to allow airflow—up to 6 inches (15 cm) deep that is covered by 2-by-6-inch (5 × 15 cm) boards with ⅜-inch (9.5 mm) spaces in between. A blower in the plenum blows upward through the spaces between the boards. If you take this approach, make sure the airholes in the PVC pipes are not directly below the open spaces between the boards, because the organic material can fall into the holes and plug them up.

Feeding Schedule

Your worms need to be fed regularly, but *they* should set the schedule, not *you*. Of course, worms don't keep a calendar, but the point is that it's very important to wait until the feed has been consumed before adding more. Worm farmers who say they feed their worms X amount daily are actually telling you a daily average, which can be figured by dividing the total amount of feed provided per week by 7. In most cases they feed their worms only every two or three days, or sometimes only once a week.

Overfeeding is one of the top problems with worm beds—so don't do it! Remember, if you add too much feed, your worm bed may become anaerobic (oxygen-deprived), too acidic, or overheated.

Managing Feedstocks on a Massive Scale

Sam Sayegh moved from Syria to California in 1989. He and his brothers became financially successful in the grocery store business. In 2012 Sam bought a 500-acre (200 ha) piece of land, named it Becky's Worm Farm, and started raising animals. He started a vermicomposting operation on the farm in January 2017 with his partner Juan Garcia.

Sam has plans for five phases of development; by 2019 the farm will have 15 windrows that are 8 feet (2.4 m) wide and 1,500 feet (455 m) long. When I visited the site in July 2017, Sam and Juan had 700-foot-long (215 m) windrows and planned to add 800 feet (245 m) to each of them. Sam started the windrows with 4,000 pounds (1,815 kg) of *Eisenia fetida* earthworms. He calculated that he will eventually have 2 pounds of worms per cubic foot (0.9 kg per 0.03 m³). The first windrow was established in March 2017. In one year the windrow will be 2 to 3 feet (61–91 cm) high.

Before Sam and Juan began feeding the worms, they experimented with different manures from the animals that are raised on the farm. One worm bed only received steer manure; another bed was fed a mixture of goat and sheep manure. The worms seemed more content, grew longer, and reproduced more in the steer manure, so they decided to go with that feedstock.

Juan applies food 5 to 6 inches (13–15 cm) deep every three weeks. They feed the worms steer manure that has been stockpiled for a year. To apply feed to a windrow, they use a Gehl forage wagon pulled by a tractor (see figure 7.8). The wagon has beaters and augers in combination with a floor apron to manage the feed material. A belt driven by a sprocket shoots manure out the side onto the windrow.

Each windrow is watered once a day for 30 minutes. They can water two windrows at a time using collected rainwater. They try not to waste water and have no runoff/leachate.

FIGURE 7.8. Becky's Worm Farm in California uses a Gehl feed wagon with an auger inside to distribute feedstock to worms in windrows that are several hundred feet long.

It is commonly stated that vermicomposting earthworms will consume 50 to 100 percent of their body weight per day, though this figure is primarily based on vegetable- and fruit-based feedstocks, which consist of more than 80 percent water by weight and therefore have less food value. Operators of large-scale dairy manure vermicomposting operations report that their feeding rate is 25 percent to 35 percent of a worm's body weight per day. And manure solids that have been squeezed through a solids separator are denser, lower in moisture, and higher in food value.

Some worm farmers who use continuous-flow reactors place marbles on the surface of the bed after feeding and see how long it takes for them to come out of the bottom of the bin. They report that it takes 40 to 60 days for the marbles to pass through the bin. This way, they know how long it takes for feed to be consumed and ready for harvesting from the bottom of the bin.

Overfeeding is particularly a problem for folks who keep a small worm bin for their kitchen food scraps. They buy a pound of earthworms and maybe a 14-gallon (53 L) worm bin. Because they are excited about trying out vermicomposting and keeping food scraps out of the landfill, they start loading up their worm bin with as much food waste as possible, assuming that the 1,000 or so worms in it are eager to gobble up all of the bits. The worms don't eat as quickly as anticipated, anaerobic pockets arise, the bin gets smelly, and the ensuing chain reaction causes conditions in the bin to go downhill, eventually resulting in the worms' death.

In the scenario I just described, the worm bin owners have very little money and time invested, and they can easily start over. However, similar incidents have happened with larger-scale vermi-operations, where there is considerably more capital and effort on the line, so take this warning seriously and proceed cautiously with feeding your earthworms. Always observe the bed before you add more feed, and wait until the last feeding is consumed before you apply more food to your worm bed.

A good indicator that a worm bed is ready for more food is what's called the pool table effect. When you first apply feed to a worm bed, there is an uneven texture to the surface due to chunks in the feedstock. When the worms have finished working through the material, the surface will be smooth like the surface of a pool table.

Applying Feedstocks to the Bed

At many vermi-operations the workers apply feed using shovels and rakes. They shovel the feed onto the worm bed and then rake it out evenly over the surface. This is the most basic and labor-intensive method of feeding.

FIGURE 7.9. Gantry feeders ride the top edges of continuous flow-through bins at PUR VER in Belgium to distribute feed to the worms below. *Photograph courtesy of Alexandre Meire.*

There are some less labor-intensive methods for applying feedstocks, which may or may not be applicable depending on the type of vermi-system you have. For example, people who have windrows often feed their worms using a manure spreader, or they approach from the side with a skid steer or front-end loader.

Automatic feeding machines used for different types of livestock can be adapted for vermicomposting. Gantry- or traction-type feeding machines can be set to continuously spread a certain amount of feed evenly. For example, some large-scale vermicomposting operations use a gantry-rail-type feed skip to apply feedstock to their continuous flow-through reactors. It moves on wheels across the top sides of the bin, uniformly delivering the feed, and it is a great labor saver.

Mass Wiggle in California feeds worms using a sand shooter attached to a Caterpillar skid steer. (See the Mass Wiggle profile in chapter 10.) As I mentioned in chapter 5, a trench-style vermi-operation used a manure spreader pulled by a tractor to feed hog manure to worms. The equipment straddled the trench and applied manure an inch (2.5 cm) deep.

I once visited a large-scale vermicomposting operation that feeds their worms three times per week and harvests castings from continuous flow-through bins twice a week. They maintain 2.3 to 2.5 pounds of worms per square foot of surface area in their worm bins (1.0–1.1 kg per 0.09 m^2).

The mantra of successful vermicast producers is, "A consistent process yields a consistent quality product." It may take several years of experimenting to develop the feedstock recipe and method of feeding that is the most suitable for your vermi-operation. Once you perfect a recipe and technique, apply it consistently.

CHAPTER 8

Monitoring Your Worm Bed Environment

Once you've set up your vermicomposting system—whether it's a windrow, pile, or some type of bin—you must closely monitor it to ensure that the conditions remain hospitable for earthworms and microorganisms. Check your beds *at least* once daily if you're running a commercial enterprise. Conditions can change suddenly, and you could lose your whole earthworm population if you do not identify problems early on and act quickly to rectify them.

When it comes to monitoring, a common mistake that many worm farmers make is to focus on the equipment (primarily the worm bins) and neglect to understand the importance of earthworm husbandry. Earthworms are the drivers of your vermicomposting operation. If they don't thrive, your business will collapse. You will lose the money you initially invested in worms and be forced to buy a new batch. Having an expensive, sophisticated vermi-system is not enough to keep your worms alive—you need to diligently monitor your entire operation and troubleshoot problems as they arise.

The task of monitoring involves keeping track of numerous environmental and biological conditions, which I explore in this chapter.

The Daily Inspection

There are several characteristics you should look for on a daily basis that indicate you have a healthy worm bin environment and your worms are thriving:

- It smells earthy like the forest, without bad odors.

133

- The worms are actively eating their feed (if they are clustered together on the upper sides or trying to get out, something is wrong).
- The contents are damp but not soggy.
- The worms should have moist, glistening skin.

Eyeing the Whole Bed

Every earthworm in your windrow or bin is important, so you need to be able to visually examine the entire bed. I touched on this point in the "Accessing the Worm Bed" section of chapter 5, but it bears repeating because I have frequently seen the issue ignored.

Be realistic about your physical limitations. If you are not very tall or agile, do you really think you can lean over an 8-foot-wide (2.4 m) bed and closely examine the conditions in the center? Consider the height of your system as well. CFTs are raised off the floor with the bed at mid-torso height, making the task of leaning over to see the center that much more challenging. Don't put a CFT against a wall; doing so will prevent you from accessing the far side of the bin. A shallow worm bed, on the other hand, can be placed on the ground against a wall because you can more easily lean across it to check conditions on the far side.

If accessing your worm beds requires getting on a ladder, you should reflect on whether that might pose a problem for you and your operation. Can you physically withstand climbing up and down a ladder throughout the day, often carrying equipment? If something goes wrong, how will you fix the problem in the hard-to-reach higher bins? And what about worker safety and liability?

Developing a setup that allows thorough examination is extremely important. Conditions on the outer edge can vary from those in the middle, and small areas of concern can develop at random locations throughout your bed. Even if your bed looks fine at first glance, you may discover upon closer examination that food is not being consumed in one spot or that a particular area has become too dry.

Moisture Control

Moisture is the most important environmental factor to monitor and control. To ensure that the moisture in your bin is maintained at about 80 percent by weight, visually inspect the bed to see if it looks moist and the worms have glistening skin. Just by looking, you will be able to see if the bedding or the worms have begun to dry out. You can also test for moisture content by grabbing a handful of material and squeezing it in your hand. If a significant amount of water drips out, it is too wet. If you open your hand and the material crumbles,

Earthworm Population Sampling

You should take a population sampling of your earthworm bed weekly—on the same day each week—to estimate the number of worms in your beds and to document fluctuations in worm population over time. This practice will help you identify potential problems and ensure that your husbandry practices are allowing your worms to thrive.

Your goal is to determine the number of earthworms in various 6-inch-square (15 × 15 cm) areas throughout your worm beds. Most of your worms will be in the top 4 inches (10 cm) of the bed,

so you should remove samples from a section that is 6 inches by 6 inches by 4 inches deep. Using your hand, reach 4 inches deep into the bed and remove a 6-inch-square area of organic material. Place the handfuls of material into a clean plastic container and count the number of earthworms, including cocoons. Keep records! Note the date, time, location in the bed, and numbers of adults, juveniles, and cocoons. Return the sample to the bed, remove a sample from another location, and repeat the procedure.

it is too dry. You'll know the moisture content is just right if you're able to squeeze out just a drop or two of water.

You can purchase a moisture meter to gauge the moisture content of your beds more precisely. Keep in mind, these are sensitive pieces of electronic equipment that must be handled with care and stored properly. If your meter is out of calibration, you may have to send it back to the manufacturer to be recalibrated, depending on whether or not the model you purchase allows you to do so yourself.

To get started on the right foot, as described in chapter 6, soak bedding in a bucket of water first so that it is approximately the right moisture content before you add it to a bin or bed. Do not add dry bedding.

Remember, the only safe way to apply water to a bed or bin is as a mist or fine spray. If you pour water into your bin or pile, you won't be able to control how much it soaks into a given area. It is too easy to saturate part or all of the bedding, filling up pore spaces with water and crowding out oxygen.

Ideally, your bed should be the wettest at the surface and get increasingly dry the deeper you go. Aim for a moisture level of 40 to 50 percent per volume at the very bottom. To achieve this, just lightly spray the top where most of the worms will be.

If rain is expected, some people cover their outdoor windrows with tarps or old carpet, but most do not because it's such a time-consuming task (many windrows are over 100 feet/30 m long). Folks with outdoor CFTs with no shelter or under a roof with open sides might cover the bins with lightweight tin or some other protection to prevent rain from falling directly on their worm beds.

Moisture level is particularly tricky to maintain in CFTs. If the moisture is not evenly distributed and parts of the bin become too dry, you run the risk of material falling through the grates at the bottom of the bin.

Water your worm beds with frequent, light applications, according to the conditions of the bed. Moisture levels will be affected by the seasonal climate and daily weather, so always be aware of these conditions. Avoid heavy, infrequent watering, and never pour water into the bin or bed (even if you have seen people do it on YouTube!). Try to keep the water close to the surface of the bed (in the top 4 to 6 inches/10–15 cm), where earthworm activity is greatest. If you overwater, your worms will follow the water deeper into the bed and breed there. You'll know you are overwatering if you find cocoons at the bottom of your continuous flow-through reactor.

It is also important to manage water runoff on your site to prevent water pollution. If you are located in the US and fall under composting regulations, you will have to meet the requirements of the Clean Water Act and the National Pollutant Discharge Elimination System. Keep raw organic materials and active compost and vermicompost piles downstream from finished vermicast.

Temperature Control

Ideally your worm beds should be kept at a temperature between 60 and 80°F (16–27°C), give or take 5°F (2.8°C). Although your worms will tolerate a range

FIGURE 8.1. These temperature dataloggers are monitoring the bed temperature of this continuous flow-through bin and the ambient air outside the bed.

of 50 to 90°F (10–32°C), they may become less productive as the temperature swings toward the extremes. Nevertheless, some worm farmers have told me that their worms thrive when it is 90°F in their beds, so a lot depends on how the bins are managed. Relative humidity has an influence as well.

Test the temperature of your worm bed and the ambient air separately. Do not assume the temperature inside your worm bed is the same as the air outside of it. Your choice of thermometer will depend on the depth of your beds: a soil thermometer will test up to 8 inches (20 cm) deep and costs about $12; a backyard compost thermometer has a 20-inch (51 cm) stem and costs around $20; and a heavy-duty compost temperature probe is 36 inches (91 cm) in length and costs approximately $120.

Cooling Worm Beds in Hot Weather

For best results, locate your outdoor worm bed in the shade. That said, I have seen plenty of worm windrows, piles, and bins in the full sun that operate just fine.

The following are some suggestions for keeping outdoor worm bins cool in hot conditions:

Reflect solar rays by painting the lids and sides of your vermi-system white, or by attaching a radiant barrier reflective fabric such as Reflectix, Silver Shade Mesh, or CoolTarp.

Provide a buffer between the outside heat and the inner temperature of the worm bin by insulating the sides with rigid insulation, hay, or straw.

Build a wood or PVC-pipe frame around the bin, and attach reflective paneling to it.

Dig a pit, and place most of the worm bed in the ground in a shaded area. It will be cooler underground, so the worms can head deeper into the bed when it is extremely hot.

Place some type of fabric on top of the bed, such as burlap bags, pillowcases, or pieces of a bedsheet. Mist the fabric to keep it moist and create an evaporative-cooling effect (like the swamp coolers that are used to cool buildings in the desert). If you have access to electricity, you can speed up the evaporative cooling by using a fan to blow air across the moist fabric.

Use freezer packs or frozen water bottles to cool the beds. Wrapping them in fabric will slow down the melting process, and condensation will keep the material moist.

In hot weather add thinner layers of feed to minimize the heat generated by decomposing food. As I mentioned earlier, you can also limit feed to certain

FIGURE 8.2. Shade cloth keeps sunlight and rain off of these worm bins in Chile.

sections of the bin so that there are cooler areas where the worms can go to escape decomposing, heat-conveying food.

Your instinct may be to cool down the bin by adding water, but that can actually cause the bin to heat up because water will fill in the air pockets in the bedding, bringing about anaerobic conditions that retain ammonia and alcohols and create acidic conditions that can kill your worms. Instead of spraying water to soak the bedding, give it a fine mist—and, if possible, set up a fan to blow air across the top of the bed after doing so.

Blooms of springtails can indicate that it is too hot in the bed for your worms. If you see them appear, you can water lightly in morning, again at noon, and then at the end of the day. In the evening after the bed has been watered, the worms will come to the surface to cool off and then burrow back into the bed at sunrise.

Keeping Beds Warm During Winter

It is usually easier overall to keep worm bins at a comfortable temperature during winter than in summer heat. But unless your worm bins are indoors in a heated space, it's important to figure out which methods will work best to keep your worms thriving during cold periods. As mentioned in chapter 7, you can add extra feed to help raise the internal temperature of the bin. Be careful not to overfeed—just apply a slightly thicker layer of food than you would during summer months. Alternatively, you can apply 1 or 2 quarts (about 1–2 L) of raw livestock manure or hay in various locations on the worm bed. These areas will be warmer, and the worms will congregate around them to feed.

To insulate the inside of worm bins, some folks place dead tree leaves on the surface of the bed. Others place blankets, rugs, cardboard, vacuum cleaner bags, or carpet pieces on top of their beds. Be careful with these types of added insulation. Make sure that air can still flow to the worms, and be aware that other types of animals (with sharp teeth!) may take shelter in the worm bin under the insulating layer.

It's also possible to insulate the outside of your bins, depending on their shape. For my 4-by-4-foot (1.2 × 1.2 m) produce bins, I use duct tape to attach 2-inch-thick (5 cm) rigid insulation sheathing to the outer sides. I leave holes on the upper sides of the bins uncovered so that plenty of air passes through, but I always keep a tarp on top to hold in moisture. Many other worm farmers place hay or straw bales against the sides of their bins to provide further insulation.

If you have a source of electricity available, even more options open up. You can place soil heating cables on top of your bins, use roof heat cables, or put seedling heat mats beneath your bins.

My fellow Master Gardener Anne Blanchard helped her neighbors across the street set up a flow-through OSCR wooden worm bin that measures 4 feet long by 3 feet wide by 3 feet tall (1.2 × 0.9 × 0.9 m), located outdoors, exposed to the elements. When we experienced record-setting low temperatures last winter, they collected several Christmas trees that neighbors had put on the curb for recycling and placed the trees around the worm bin to insulate it from the cold temperatures and shield against blustery wind gusts.

In milder climates such as California, people are more likely to keep worm bins outdoors. During the winter, some worm farmers with continuous flow-through bins will add a thicker layer of feed and then lay cardboard on top to insulate the bin. Others will drape canvas over the worm bin all the way to the floor and place lights under the bin to generate heat, which is held in by the canvas. Note that using lights in this way does create a risk of fire. Another worm farmer I met covers his indoor CFT with a pool cover, rather than canvas (and he does not use lights).

If your worm bins are inside a building, it is not necessary to heat the building to 70°F (21°C) or more. If temperatures regularly drop below freezing, this would cost you a fortune! A worm farmer in the northern United States heats his building to only 40 to 45°F (4.4–7.2°C) during winter, and his worm bins stay at 80 to 85°F (27–29°C). Of course, it is his worm husbandry skills that account for much of the difference in temperature.

A few worm farmers, such as Michigan-based Gavin Newton, have even built a "room within a room" by constructing a frame with insulated walls around their worm bins. Others have built tents around their worm bins, made from frost protection blankets, to trap radiant heat.

Lighting

Lights can be used to prevent worms from leaving their beds. Earthworms are prone to crawl away under a variety of conditions, including:

- Rain
- Fog (excess moisture in the air)
- Barometric pressure changes
- Intense wind
- Dryness in bin
- Lack of food
- High acidity in bin
- Desirable feedstock right next to bin

Remember, when it rains, earthworms are naturally inclined to leave their homes to move about in the moisture, so you may find that some of yours crawled off during a rainstorm. Half a dozen worms inside my Worm Barn usually crawl out of their bin during hurricanes or if it rains for several days in a row. If you live in a climate with high rainfall, take extra precautions to prevent your worms from leaving the bin or beds.

As mentioned in chapter 3, earthworms are sensitive to light, so you can use artificial light to keep them contained. If it is brightly lit outside the bedding, worms will remain burrowed beneath the surface. Remember, worms don't sense red light, so avoid using red light bulbs to prevent them from escaping the bin.

Earthworm Predators or Annoyances

Earthworm predators in nature include birds, rats, snakes, carabid beetles, moles, mice, gophers, toads, and a few species of carnivorous slugs or mollusks. Centipedes will prey on worm cocoons or young earthworms, but they do not usually attack earthworms. Some species of planarian flatworms eat earthworms, but they don't typically cause problems in commercial worm beds. Here I will describe some of the animals that can interfere with a vermicomposting operation, ranging from minor pests to more serious predators.

Flies

Flies aren't a danger to earthworms, but they can be annoying to humans. Fungus gnats and humpbacked flies may live in worm bins and dash around the top of the bedding.

Fruit flies, vinegar flies, and pomace flies may also be found in worm bins. Fruit flies lay their eggs on the surface of fermenting foods or other moist, organic materials. They can lay 500 eggs, and the life cycle from egg to adult can be completed in 7 to 10 days. I encourage worm farmers to be as vigilant as possible about covering food scraps in their worm bins. If you see two fruit flies, soon you will have thousands! If food scraps are completely covered by bedding, fruit flies won't even know they are there.

The only time I've experienced annoying flies was about 20 years ago, when I had seven household worm bins set up inside a lab at the university and was picking up dining hall food waste once a week to feed to the worms. After a while I noticed little flies that I had never seen before. I looked them up on the internet and discovered that they are called trickle filter flies (or midge, moth, or sand flies). They lay eggs on wet, nutrient-rich films and are usually found at wastewater treatment plants. I realized that they were attracted to my worm bins because the dining hall food was cooked, which meant that the feedstock was full of liquid. As a result of that experience I now caution my audiences about the high liquidity of cooked food. I tell them that it's okay to put small amounts of cooked food in worm bins but to avoid adding too much.

Mites

Mites are naturally found in soils, manures, and other organic materials. All worm beds contain small populations of mites, but under certain conditions mite populations can reach extremely high levels. If worm beds are not cared for properly, acidity can build up and create conditions that allow mites to thrive. You should routinely check pH and add agricultural lime if the pH is less than 6.8.

White or Brown Mites

White or brown mites are not predaceous and will usually feed only on decaying or injured worms, but if their numbers increase significantly, these mites can eat most of the food in earthworm beds, depriving worms of needed nutrients. This increases worm growers' costs and time spent feeding worms. Mite infestations can also cause earthworms to remain deep in the beds and not come to the surface for feeding, resulting in poor growth and reproduction.

Red Mites

These are the mites that can cause the most problems in a worm bed, because they can parasitize earthworms. Red mites can prevent earthworms from feeding, and—worse—they can attach themselves to worms and suck out their blood or body fluid. Red mites can also pierce and suck fluids from egg cocoons. These creatures first appear as small white or gray clusters resembling mold; magnification will reveal clusters of juvenile red mites in various stages of development. The adult red mite, which is smaller than the white or brown mite, has a bright red, egg-shaped body with eight legs. Methods for preventing and removing mites from worm beds are discussed in the "Troubleshooting Worm Bed Conditions" section later in this chapter.

Hammerhead Worms

Since late 2015 I've received a few emails from folks concerned about seeing hammerhead worms in their worm beds. Apparently there have been alarming news articles about them—you know how fear sells more news—so some people freak out when they see these flatworms. These alarmist articles purport that the hammerheads will eat all of the earthworms in your garden and lawn.

What is a hammerhead worm? Also called the flatworm, arrow worm, or broadhead planarian, it is a land planarian of the genus *Bipalium* with a head plate that resembles that of a hammerhead shark. Several *Bipalium* species are considered invasive to the United States, but they have actually lived here for decades. If you are seeing them for the first time, it's not because they have recently arrived in the US; it's likely due to recent rain, which entices them to move about, just like earthworms.

Although some hammerhead worm species will consume earthworms, other species eat mollusks, slugs, and insect larvae instead. Hammerhead worms do not pose a serious threat to earthworm populations, so don't worry if you spot one in your yard or garden. If you find one in your worm bed, remove it.

For those who are still worried about flatworms/hammerhead worms: There are no proven chemical methods for controlling flatworms. To prevent the introduction of flatworms to your property, when you buy or bring in plants, examine the soil in the pots or trays to see if there are flatworms. You should also ensure that flatworms are not entering open bags of compost or growing media. Like composting earthworms, flatworms live on the soil surface and seek damp places, such as under logs, rocks, plant containers, and plastic sheeting. Check these areas regularly to see if there are any flatworms.

Centipedes

Once in a while someone contacts me about some type of "pede" in their worm bin. Millipedes will leave worms alone, but centipedes can kill worms, so you should remove them from your worm bin. Be careful about grabbing them—they can bite you with a pair of poison fangs. You can don a glove and crush them between your thumb and forefinger, and then leave the dead carcasses in the worm bin to be devoured by other critters.

Ants

Occasionally ants may be a problem or annoyance for worm growers, as some species will feed on cocoons and small worms. Ants are attracted to high-concentrate feed in worm beds. Some people use physical barriers around worm beds to keep ants out: One method is to apply a 1-inch (2.5 cm) paste of Vaseline around the legs or upper sides of a worm bin; another approach is to put the legs of a worm bin inside buckets containing mineral oil. Sprinkling diatomaceous earth around the outside of worm bins will disrupt the ant trail. There are also chemical baits and sprays that will kill ants. If you face a serious infestation and decide to use one of these methods, apply them only outside the worm bins, because these substances can also be harmful to worms.

At a presentation at the 2017 NC State Vermiculture Conference, Maria Rodriguez, who owns Byoearth in Guatemala, described techniques she uses to discourage ants from entering worm bins. Maria uses two types of food around the inside edges of bins to deter ants: onions and cooked rice, which develops fungus after four days. Ants don't like limestone, so Maria also uses that to repel ants by putting it around the bottoms of the legs of her raised worm bed.

Several years ago we were feeding hog manure to worms inside the 8-by-5-foot (2.4 × 1.5 m) continuous flow-through bin (CFT) in the Worm Barn at the Compost Learning Lab. Here in North Carolina, not only do we have little black ants, but we have fire ants as well. These aggressive ants were prolific in several areas of my lab at that time, and some of them beelined for the hog manure in the CFT bin. A steady stream of ants came from various nests outside the barn, and despite trying both barrier methods—Vaseline and mineral oil—I could not deter them. The ants would continuously march into the oil or Vaseline until their dead carcasses formed handy bridges for their comrades to trek over. I finally consulted an ant specialist in the campus entomology department. He tried a few approaches and eventually had to resort to applying insecticides outside the Worm Barn. That did kill the ants, and they have not returned since.

Black Soldier Fly Larvae

Unlike many other types of flies and maggots, black soldier fly larvae (BSFL) are not considered pests or disease carriers. The larvae are a dirty whitish color, about 1 inch long and ¼ inch wide (25 × 6 mm), with 11 segments plus a protruding head. They are organic material decomposers just like earthworms used for vermicomposting. They have voracious appetites, and it's pretty amazing to watch the larvae tear into whatever you feed them. Based on my observations, I call them the piranhas of the insect world. You know I love earthworms, but when it comes to feedstock consumption, black soldier fly larvae put worms to shame. Therein lies the problem if black soldier fly larvae invade your worm bed—they will outcompete the worms for food, and your worms will retreat deeper into the bedding and won't be able to consume what you are feeding them. So if you discover black soldier fly larvae in your worm bed, it's a good idea to remove them. You can put them in your compost pile or set up a separate bin solely for these amazing decomposers.

Troubleshooting Worm Bed Conditions

During your daily inspection of your worm beds, it's also smart to watch for conditions that indicate something is wrong. Here are signs that something is amiss, indicating a need to troubleshoot your vermi-systems:

- Worms are dying or leaving the bed or bin.
- Mold is forming.
- Flies are swarming around.
- Bedding is drying out.
- Liquid is collecting at the bottom of the bed or dripping out.
- The bin smells bad.

As you've learned, if your worms are crawling away or dying, there are several possible causes: extreme temperatures, inadequate oxygen, lack of food, or excessive moisture or dryness. Another crucial factor could be elevated levels of salts or ammonia in the feed. Even tiny amounts of ammonia gas, which can form from fresh manure or sour grains, have the ability to kill earthworms. Gas is especially perilous in enclosed worm bins.

It is normal to see a little bit of mold in a worm bin, but if you see a lot of mold, it usually indicates a problem. Overfeeding is the primary cause of an overabundance of fungal growth. The second most common cause is acidity. Low-pH conditions are created by a glut of food or an overload of a specific kind of food, as explained later in this chapter.

If your bin is attracting flies, it could be due to overfeeding or exposed food scraps. Reduce the quantity of feed you are offering your worms; if you are feeding food wastes, try covering them with bedding. These simple steps can save you a lot of work and frustration.

If liquid is leaking out of the bottom of your bed or bin, you have added too much moisture. There could be poor ventilation, or your feed may contain excess water. To remedy the situation, remove overly moist food scraps, add dry bedding, and take the lid off and let the bin air out.

Your worm bin should smell earthy like healthy soil. If it has an unpleasant odor, you are not maintaining the bin correctly. The smell is likely due to inadequate ventilation, excessive moisture, and too much food. If you are adding table scraps and not covering them with bedding, they could also produce odors as they decompose. To eliminate odors, mix in dry bedding and leave the lid off for a while, allowing your system to air out.

pH

Check the pH of your worm beds periodically, using either a pH meter or litmus paper. If you are using litmus paper, saturate the strip with moisture in the bedding and the paper will change colors to indicate the pH level. Determine the pH level by comparing the color of the litmus paper to the color guide that came with it. The lower the pH reading, the more acid your bed is, and the higher the reading, the more alkaline it is. An acceptable pH range for a worm bed is 6.0 to 7.4.

It may seem that a 1-point difference in pH wouldn't matter, but the pH scale is logarithmic, so for each 1-point change, the acidity or alkalinity of the

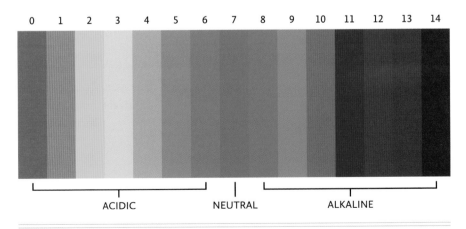

FIGURE 8.3. pH is expressed on a scale from 1 to 14. A pH of 7 is neutral; values below are acidic, and those above are alkaline.

material is actually increased or reduced by a factor of 10. For example, vermicast with a pH of 6 is 10 times more acidic than vermicast with a pH of 7, and it is 100 times more acidic than vermicast with a pH of 8.

Ammonia, a form of nitrogen gas that is toxic to worms at low concentrations (0.5 ppm), will form if the pH is above 8.0, so try to keep the pH of the bin below 7.5. Active compost or manure can create ammonia gas, but the gas will dissipate if the vermi-system is kept in an open container.

Bedding Too Dry

If the worm bed is too dry, it can harm your worms. Worms need moist skin to breathe, so if they are in dry conditions they must exude body fluids to moisten their skin. This process will weaken your worms and cause them to eat and reproduce more slowly. Eventually, they will die.

Red Mite Infestation

Periodically a worm farmer will leave me a message exclaiming that their worm bed is overrun with bright red mites that are outcompeting their worms for feed.

It is common for several distinct species of mites to be present in worm beds. They are members of the soil food web that aid in breaking down food. However, if your bed has an overabundance of red mites that is interfering with your worm farming operation, it means you are not properly maintaining the bed. Here are some conditions in worm beds that are usually associated with high mite populations:

TOO MUCH MOISTURE. Prevent overly wet beds by monitoring the moisture level, altering your watering schedule, and improving drainage.

OVERFEEDING. Too much food can cause fermentation and lower the pH of the worm bed. Modify your feedstock application rates so that the food is completely consumed within a few days. Adjust your feeding schedule as the seasons and temperatures change. Maintain a neutral pH of 7 in your worm beds.

EXCESSIVELY WET OR FLESHY FEED. Limit the feeding of fruits and vegetables with high moisture contents.

If the top of your worm bed is swarming with bright red mites and your worms have retreated into the bedding below them, you might have a mite problem. Mite removal methods range from simpler to more intensive approaches:

Method 1. Remove the lid or cover from your worm bed, and subject it to light for several hours. Don't feed or water the bin. The mites may leave the bin because the environment is no longer favorable to them.

Method 2. Moisten newspapers or burlap bags, and place them on top of the beds. After mites accumulate on the moist paper or burlap, remove it. Keep repeating this process until most of the mites are gone.

Method 3. Place slices of watermelon or cantaloupe on the surface of the worm bed. Mites will climb aboard to devour the sweet fruit. Remove the mite-laden melons and either drop them in water, bury them, or put them in your compost pile.

Method 4. Some worm farmers control mites by watering the bed heavily, being careful not to flood it. Mites will rise to the surface, and worms will stay below the surface. The farmers then use a handheld propane torch to scorch the top of the worm bed and kill the mites. I do not like this method, but some people find it effective. This process can be repeated several times, at three-day intervals, if needed. Keep in mind that this procedure could be a fire risk.

Method 5. Use agricultural sulfur dust to kill the mites. Add water to the surface of the worm bed, prompting the mites to emerge, and apply sulfur at a ratio of 1/16 ounce per square foot (1.8 g per 0.09 m²) of bed surface. Sulfur should not harm the worms, although after a period of time it could increase the acidity of the bed and reduce the population of earthworms.

Deformed Worms

If your worms look deformed and lumpy, they may have what is called protein poisoning, also known as sour crop. As the condition progresses, they may eventually resemble a string of pearls; parts of the worm can break off, and they will die when their intestines rupture. Protein poisoning occurs when the worm feed doesn't have the proper balance of protein, carbohydrates, and enzymes. Feed that is too acidic will ferment in the worm's intestines and produce gases, which build up and eventually burst the intestine.

If you have worms afflicted with sour crop, you won't be able to save them. These worms will die, but fortunately the other worms cannot catch it. Do away with the afflicted worms, and fix the conditions in the bin. Remove uneaten food that is fermenting in the bin, and raise the pH of the bedding by lightly sprinkling an alkaline substance such as calcium carbonate, crushed limestone, or ground eggshells.

This affliction is caused by a combination of overfeeding and acidic conditions. It can happen when you add citrus fruits or too many coffee grounds to your bin. However, it seems that the majority of reports of protein poisoning are from growers who feed grains to their worms, such as cornmeal, wheat barley, or rice products.

As with so many vermicomposting problems, one way to prevent this is to wait until your worms have consumed all of their previous meal before serving them more to eat.

Recording Observations

It is vital that you and your staff keep timely, accurate records of everything you do and observe. The effects of something you did today likely won't be visible for several weeks or months, and by then you will not remember exactly what you did in the first place to bring about the effects. Therefore, you need to write things down! On a daily basis, record the date, time, location, what was done, how things looked, and anything else that may be of importance—an unusual weather event, a time change, what have you.

Don't write notes on slips of paper that will get lost or be hard to decipher. Instead, use mobile computing devices or keep weatherproof logs at different locations throughout your operation so you and your workers can easily record all activities that are being carried out. Maintain separate logs at each of the following locations:

- Feedstock receiving
- Feedstock processing
- Pre-composting bins
- Worm bins
- Harvesters
- Bagging
- Shipping
- Equipment fueling and repairs
- Office

If you prefer recording on paper, design a blank template and then make copies. If you feel comfortable using computers, there are several options for keeping records. Google Sheets is a free online spreadsheet application that allows you to input data from your phone, tablet, or computer and then saves the information automatically. Windows and Android operating systems have an Excel spreadsheet application; Apple's version is called Numbers. All of these applications are fairly intuitive to use. I am not a computer whiz, yet I manage to use Google Sheets or Excel almost daily.

Your record keeping—whether digital or on paper—can be simple. Table 8.1 shows a sample harvest log.

If you are pre-composting your organic materials, thorough record keeping is particularly important. Pre-composting is as much an art as a science. There is no universal recipe that you can download, so you will have to develop

TABLE 8.1. Sample Harvest Log

Bed #	Date of Harvest	Amount Harvested
1	3/15/17	X pounds or kilograms
2	3/15/17	X pounds or kilograms
3	3/16/17	X pounds or kilograms
4	3/16/17	X pounds or kilograms
Total Harvested		**X pounds or kilograms**

your own recipe through a continued process of adjusting and tweaking. It is imperative that you write down everything you do when you modify your recipe so that you can remember the changes you made when the effects manifest weeks later.

Information you should record about your composting endeavors includes the following:

- Pile creation dates
- Recipes
- Turning dates
- Temperature readings, dates, and locations in the pile
- Water addition volumes and dates
- Rain events
- Date material was removed for further processing in worm bins

I have provided an example of a daily composting temperature log in table 8.2. Be aware that your state may require additional information, so you should figure out what their specific requirements are. You may want, or need, to add more detail to the log, such as the locations and depths in the pile where temperatures were taken.

If your operation is permitted by state or local authorities, you will be required to maintain records and keep them available for inspection or to submit for review. Required record keeping may include: organic waste

TABLE 8.2. Sample Daily Temperature Log

Date/Time	Outside Temperature (°F/°C)	Bin or Pile ID	Compost Temperature (°F/°C)
2/15/18; 8 AM	40/4.4	Bin #1	135/57
2/16/18; 8 AM	42/5.6	Bin #1	138/59

received from off-site, amount of vermicompost sold, and temperature records for composting piles. If you are receiving waste feedstocks, you may be required to keep a log of the dates of delivery, types of material, volumes or weights, and sources. Check with your state or local authority to figure out what specific record keeping they require.

CHAPTER 9

Harvesting and Post-Harvest Practices

N ow that you've learned about the main steps involved in starting a vermicomposting operation—from crafting a business plan to choosing a vermi-system to formulating a feedstock recipe and getting your worms to work—it is time to discuss the reward for all your efforts: methods of harvesting and processing your products. In this chapter I first address how to remove worms from your vermi-system so you can sell them or use them to start new windrows or bins. I then move on to instructions for harvesting vermicast.

If you have a farm or volunteer in a community garden, you may want to use the vermicast on your site to grow crops. If you went into the vermi-biz to make money, then you will be selling vermicast or worms. That involves packaging and shipping earthworms, testing your vermicast, storing it, bagging and labeling vermicast, and making tea/aqueous extracts.

All of these steps will be guided by your marketing and business plans. The site plan you developed shows where harvested worms and vermicast go to be prepared for markets. It also designates where to safely store vermicast away from the elements.

I devote quite a bit of attention to testing vermicast so you will understand how to take samples, where to submit them for laboratory testing, and how to interpret the lab results.

If you decide to sell vermicast in bags, please understand that labeling the bag is a complicated process, so read this section of the chapter carefully. Every state has its own regulations concerning what you can print on your bag, so be sure to find out what the rules are for your state—and every state where you plan to sell your products.

I finish the chapter by discussing vermicast tea. To thoroughly cover this topic, it would take a separate book. There still exists some disagreement about how to make and use tea, in addition to its effectiveness. So I touch on some aspects of vermicast tea brewing and use that I think are important to consider.

Manual Earthworm Removal

Table harvesting and *light separation* are two different names for a common earthworm harvesting technique. To use this method, start by placing a table next to your worm bin. Cover the table with a waterproof plastic sheet, such as an old shower curtain, and set up containers near the table. Use a digging fork or shovel to remove the top 4 to 6 inches (10–15 cm) of material, where most of the worms reside, from your bed, and place the material on top of the plastic. Shine a bright light on the pile of harvested worms to encourage them to move deeper. Remove the organic material and put it back in the bin for further processing. As the worms move away from the light, they will eventually form a solid mass, which you can then place in the containers.

Vertical separation is a method that lures worms to move upward toward food. It helps if you do not feed the worms for a week or two before using this technique. Place one of the following on top of the worm bed: a piece of mesh, a plastic bakery tray, a bulb crate, or a box with wooden sides and a mesh bottom. Put enticing food, such as melon, pumpkin, or Purina Worm Chow, on the mesh; the worms will crawl through it to get to the food. Then you can lift the tray or mesh off the bed, and voilà, you've got your worms!

Vermicast Harvesting

If you plan to harvest your vermicast by hand, you'll need harvesting boxes. Construct two boxes out of wood 24 inches by 18 inches by 6 inches deep (61 × 46 × 15 cm). Attach ¼-inch (6 mm) screen to the bottom of one box and ⅛-inch (3 mm) screen to the underside of the other. Scoop out material from your worm bed and place it in the first box on top of the ¼-inch screen. Place a wooden box or plastic storage tote underneath the box to catch whatever falls through, and then shake the box so that the organic material will pass through the screen, leaving the worms behind. It helps if you thump the side of the box with the flat of your hand so that the worms jump up from the screen instead of crawling through. Return the worms to the bed or place them in a plastic mortar tray if you plan to sell them or start a new bed. Next, dump the material in the container into the box with the ⅛-inch screen. Shake and thump the box

FIGURE 9.1. Doug Tompkins designed this beautiful pentagon-shaped worm bin in Chile based on the sideways separation method of harvesting. Food entices worms to move laterally into each section of the bin.

to make the finer vermicast fall through; cocoons and unprocessed material will remain on top of the screen. Return the cocoons to the worm bed, or place them in a new bed or breeder bin.

Sideways separation is a method that entices worms to move horizontally in the pile or bed toward fresh feed. This is the concept behind the wedge system described in chapter 5, and it has also been used in other scenarios. It works for a worm bed of any length. Feed the worms until their bed is 18 to 24 inches (46–61 cm) deep. Then abstain from feeding the worms for one or two weeks. Next, stop watering the pile for a couple of days. After that, set up a new area adjacent to the pile, or start to build out a wedge if you are using the wedge method. Add moist bedding and feedstock to the new area, and the worms will move laterally, attracted to the moisture and fresh food. The worms will not make an immediate beeline to the new food, so be patient during the next few weeks (or even months), and check periodically to see whether any worms are still in the area of the bed that is not being fed. Some will remain awhile, reingesting their castings, before moving into the bed with fresh food. When it appears that there are few remaining worms in the original bed, you can harvest the vermicast.

When I taught in the Dominican Republic, the farmers had only 2 pounds (0.9 kg) of worms to add to their 10-foot-long (3 m) bed constructed from concrete blocks (see figure 5.4). I suggested they use some additional concrete blocks to wall off a 2-foot-square (0.2 m²) section and put the worms there. After that part filled with vermicast, they could set up another temporary wall and add moist bedding and food to entice the worms to move into that new area.

Harvesting Windrows or Bins

To harvest vermicast from worm windrows or bins, growers use manual or mechanical methods, and both involve using a trommel (cylindrical) or flat shaking screen. Since most of the worms will be in the top 4 to 6 inches (10–15 cm) of material, if workers remove the top 6 to 8 inches (15–20 cm) of the windrow or bin, using pitchforks or a front-end loader, they will capture most of the worms. They place this worm-rich material in a new windrow. Then they use a shovel or front-end loader to pick up the rest of the organic material and empty it onto a screening device. If the vermicast is too moist for screening, it can be rototilled or turned with a shovel or a loader to aerate the pile and then left to allow moisture to evaporate for a day or more.

Harvesting Continuous Flow-Through Bins

With continuous flow-through bins, you can harvest the vermicast without disturbing the worms or the bed. Most of the worms will be in the top 4 to 6

FIGURE 9.2. This trommel screen on wheels can be moved over a windrow to harvest vermicast and worms.

inches (10–15 cm) of the bed, and the rest of your CFT will be almost full of organic material (feedstocks, bedding, and vermicast). There are various methods for harvesting these beds that involve prompting a layer of finished vermicast (no more than a 1-inch/2.5 cm thickness) to fall through the grates in the bottom of the bin.

The trick is to not have *all* of the material in the bin fall through the grate at once. How you accomplish this depends on the way the bin was built.

Some people build their continuous flow-through bins using parallel metal rods rather than a crisscross grate. With this setup, you can take a four-tine cultivator rake (or something similar that fits your system), turn it upside down so the tines are facing upward, and scrape it along the underside of the grate. The tines will loosen the bottom inch or so of vermicast and make it fall through. It's important to be gentle and only use the very tips of the tines to rake the material, or else too much material will loosen and fall through.

Most continuous flow-through bins have crisscross grates like mine that are 4 inches by 2 inches (10 × 5 cm). To harvest from this type of continuous flow-through bin, folks will usually use a hand-operated crank or pulley, or a hydraulic system activated by small motors. These mechanisms pull a breaker bar or blade over the surface of the grate, shaving off 1 inch or less of vermicast.

The CFT I purchased for my Worm Barn has a breaker bar attached to a cable. The fastener that held the cable to the breaker bar was not very sturdy, so the first time we harvested the bin, we activated the motor to drag the bar across the top of the grate, and the cable snapped off! What do you do with a disconnected cable at the bottom of a worm bin that is filled with tons of material? After uttering a few choice words, you crawl under the bin, lie on your back, and try to repair the cable, which involves keeping your arms extended upward until it is excruciatingly painful—that is, if the bin has enough room for someone to lie under it. This is one reason beds are raised. During winter, it is especially pleasant to lie on the freezing concrete floor while trying to repair your cable. When you feel like your arms are about to fall off, with luck you will have been able to tie the cable in a knot, so you can finish harvesting your bin.

I hope my charming description has inspired you to use the most durable possible materials for your harvesting mechanism. I have not seen it in person, but the chain-and-sprocket harvesting system used by Germina, Inc., looks quite durable. (I describe this Turkish operation in detail in chapter 10 on page 200.)

The harvested material falls through the grate and lands on one of the following: the bare floor, a tarp, the inside of some type of strong yet lightweight container, or a conveyor. The CFT bin must be at least half full before you take

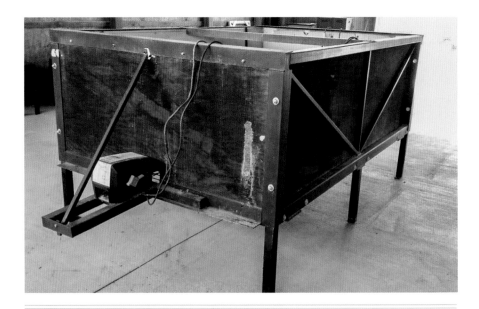

FIGURE 9.3. My continuous flow-through bin has a motor on each end to pull the breaker bar across the top of the grate for harvesting.

the first 1-inch-thick (2.5 cm) cut, to ensure that the rest of the vermicompost stays knit together and doesn't fall through all at once. Cutting usually takes place once a week, and often the vermicompost is left sitting on the floor for a day or more to allow some of the moisture to evaporate.

Instead of a tarp, some people prefer to capture vermicast on pieces of LDPE plastic (the material that is used to line ponds). This thin plastic is lightweight yet very durable, resistant to tears and punctures, and leakproof.

People use manual or mechanical methods to remove vermicast from underneath a continuous flow-through bin. They may use a shovel, broom, rake, or long-handled squeegee to scoop or scrape it out. Others install mechanical devices such as conveyors or motor-and-winch-operated manure slurry scrapers.

Screening

Some worm farmers include a screening step in the harvesting process to separate worms from vermicast, and others do it to separate fine, processed material from unprocessed larger chunks. The latter is often to meet customer demands for particular markets. How small (fine) you want the particles to be depends on what the vermicast will be used for, such as making teas or use in a soil mix. You will need a ⅛-inch (3 mm) screen to yield fine, uniform vermicast free of cocoons. A ¼-inch (6 mm) screen will allow cocoons to pass

FIGURE 9.4. A trommel can be built from bicycle rims and screen.

through along with larger, undigested particles. For some markets, vermicast producers will screen vermicast even finer, using a ⅒-by-⅒-inch (2.5 × 2.5 mm) grid.

Overs are particles that are too large to pass through the screens; they contain worms and larger pieces of undigested organic material. Worm farmers handle overs in diverse ways. Some put overs back through the composting or vermicomposting system. Others toss them in a field, where their nutrients, organic matter, and microorganisms will benefit the soil. Some vermicast manufacturers sell overs at a price lower than their vermicast. For example, one worm farmer sells his vermicast for $400 per cubic yard and his overs for $200 per cubic yard.

High-moisture vermicast will clump together and resist passing through a screen. Because of this, most worm growers spread out the vermicast on the ground or floor to allow some moisture to evaporate. When the moisture level is about 45 percent by weight, vermicast should be ready for screening.

You can screen vermicast with either a hand-built screener or a commercially manufactured device. For small operations, you can make a simple screener using hardware cloth and a wooden frame. Place it on top of a wheelbarrow or container and use a gloved hand or a shovel to push the materials through the screen.

A trommel is a rotating cylinder usually measuring 5 to 10 feet (1.5–3 m) in length and 2 to 3 feet (61–91 cm) in diameter that is elevated at one end. The

FIGURE 9.5. Vermicast is loaded into a wooden hopper, then conveyed into a double trommel. A conveyor delivers it to the bagging area at The Worm Farm in California.

sides of the cylinder are screens of varying sizes. A metal shaft runs lengthwise through the middle and is attached at one end to a hand crank or (more often) a ½- to ¾-horsepower (0.4–0.6 kW) electric motor that makes it rotate about 10 revolutions per minute. Most of the worm farms I describe in chapter 10 made their own trommel harvesters. They are all similar in appearance and did not cost much to make.

Material is added to the top end of the trommel; as it rotates, vermicast particles that are small enough will fall through the screen. The rest of the material will come out the other end of the cylinder, and the worms are caught separately. It is important that vermicast not be too moist or it will clump and be unable to pass through the screen.

If you do not have the desire or means to build your own trommel screen, you can purchase one from a manufacturer. The trommel harvester from Jet Compost Products is the most commonly purchased commercial trommel for screening vermicast or compost. Andy Dombey owns Jet and manufactures the units in Nevada. Trommel screeners sell for $2,500 to $5,000, and they are shipped all over the world. Andy created the first flat screen trommel, called the OctaScreener due to its octagonal shape with flat screens. He says the flat screens are superior to round ones because they are easier to handle, ship better, and can be replaced more readily. They are manufactured in a variety of sizes to fit different needs. The company even offers a model that harvests windrows.

The Brockwood Worm Shi*fter by Harry Hopkins is a screening unit that is selling for $2,700 plus shipping. Instead of a rotating cylinder, this device has a horizontal screen that shakes to make the material fall through. The frame reminds me of a lawn mower in that it has two big back wheels, two smaller wheels in front, and a handle so you can easily push it around to various locations on your site. The screening system separates vermicast, cocoons, and worms using a ⅓-horsepower (0.2 kW) motor to shake the screen. Harry says it can process 240 pounds (109 kg) of worms and up to 1,000 pounds (455 kg) of vermicast in an hour.

Large operations will often use a forklift to pick up harvested vermicast in a MacroBin or a loader with a bucket to pick up loose vermicast. Then they dump the vermicast into a hopper, and a conveyor transports the vermicast to a trommel screener.

Packaging and Shipping Earthworms

Vermicomposters sell worms by weight, and the pound or kilogram is the standard. If worms are sold for bait, they are usually sold by count. Growers will sell bait worms or breeding stock or bed-run worms. The following are the differences:

- Bait worms: at least 2½ inches (6.4 cm) in length and ⅛ inch (3 mm) or greater in diameter
- Breeding stock: large worms with fully developed clitella
- Bed-run: worms in all stages of development

If you plan to sell worms for bait or breeding stock, you need to sort them by hand. Count large worms as you put them into a bait container or set them aside for weighing.

Some growers will gather worms in a 5-gallon (19 L) bucket and toss them into a trommel screener to clean them. Others will entice worms to move upward through a screen to fresh feed as described at the beginning of this chapter.

Before shipping earthworms, some worm farmers sprinkle Purina Worm Chow on top of the bedding inside the box. Others warn not to put food in the bedding when you ship worms because it can heat up and kill them.

Often worm growers ship 1 pound of worms to 1 pound of organic material. Some use sphagnum peat moss and others use coconut coir as the bedding for shipping. Both work well and cost about the same, although coir is coarser. Coir may contain some sodium chloride depending on its origin, so you should test the salt level.

Don't pack the shipping bag or box full of bedding; it is important to leave space for oxygen. Place the bedding in loosely instead of packing it in tightly.

When The Worm Farm in California gets ready to ship worms, they take a shallow plastic container that weighs 2½ pounds (1.1 kg) and place 4 gallons (15 L) of peat moss in it. They set the container on a scale and add water until the weight is 12½ pounds (5.7 kg). Then they add worms until the total weight is 22½ pounds (10.2 kg).

They place the worms and peat moss in Agribon AG-50 breathable bags and seal the bags with zip ties. A bag is set on top of a paper egg carton on the bottom of the cardboard shipping box: The carton will absorb excess moisture from the bag so the box doesn't disintegrate. The worms will last one week in the bags for shipping.

Agribon AG-50 is polypropylene fabric sold as row cover to protect trees, shrubs, and crops from heavy frost damage with 50 percent light transmittance. It comes in rolls that are 83 inches wide and 50 feet long (2 × 15 m). The Worm Farm pays someone to cut the fabric and sew it to size for 1-, 2-, and 5-pound (0.5–2.3 kg) bags.

Another worm grower in California orders shipping supplies from Uline in Wisconsin. He uses a 6-by-6-inch (15 × 15 cm) box for shipping 1 pound (0.5 kg) of worms. The worm farmer punches holes in every side of each paper box. He buys wax from a craft store and melts it in a tin, then brushes this over the inside of the box to prevent moisture from breaking it down. He also purchases

FIGURE 9.6. This machine is used to slice paper egg cartons at The Worm Farm in California for shipping worms.

FIGURE 9.7. Two 5-pound (2.3 kg) bags of worms ready to be shipped to a customer.

freezer packs from Uline to put on top of the box contents to protect worms when shipping in hot weather.

Of course, you can purchase boxes that are already coated with wax. They are more expensive than uncoated boxes; coating the boxes yourself is a way to save money.

Make sure the shipping company knows that the boxes contain live creatures that need special care. Mark every box LIVE EARTHWORMS. HANDLE WITH CARE. DO NOT EXPOSE TO EXTREME HEAT OR COLD.

Be sure to ship worms only on Monday or Tuesday to help ensure they arrive at their destination by Friday. If you ship later in the week, you run the risk of the worms being stuck somewhere that is too hot or too cold over the weekend; they will have a lower chance of survival. As a courtesy to your customers, let them know when their package of worms gets shipped and when they should expect delivery.

When you ship 1 pound (0.5 kg) of worms using the United States Postal Service (USPS), you will need to dispatch it as weighing 3 pounds (1.4 kg). This is because you will have 1 pound of worms and 1 pound of peat moss or other bedding material, plus the box will weigh approximately ³⁄₁₀ pound (136 g) or more. There are eight postal zones in the US, and the postal service charges differently for shipping to each zone from your location.

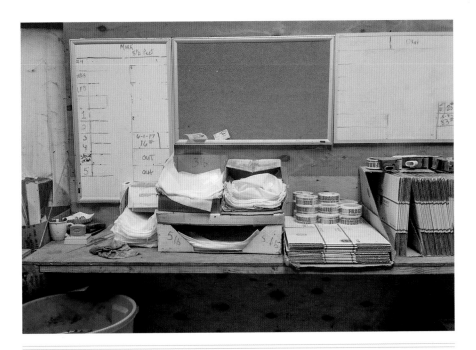

FIGURE 9.8. This workbench has supplies neatly organized and labeled for shipping worms. The whiteboards help keep track of orders and shipments.

Scores of worm growers use the USPS Click-N-Ship service. You can print your own prepaid Priority Mail shipping labels on regular paper and glue them to USPS Flat Rate boxes. The labels contain delivery confirmation numbers that track the date and time of delivery. You can request free package pickup or mail the boxes yourself. Shipping USPS two- to three-day Priority Mail will ensure that your worms won't be in transit for a prolonged period of time.

You can also ship worms by FedEx or United Parcel Service (UPS), but many worm growers have found the US Postal Service to be the better option. Contact each service and ask how they handle packages. That will help you decide which service is less expensive and more convenient than the others for you.

It is common for growers to guarantee live shipment of worms for their customers, and replace them for free if the worms are dead on arrival. It is wise to require customers to contact you within 24 hours of delivery if there are problems and they want a refund or replacement. Some worm growers will not ship to very hot or cold places because temperature extremes reduce the worms' chances of survival.

Many folks use QuickBooks to keep track of shipping and sales. This accounting software automates data entry, helps monitor expenses, and provides financial reports for running your business. Several internet stores will manage merchandise sales through QuickBooks, including PayPal and Square. For example, QuickBooks will send invoices to customers, and they can make online payments using PayPal.

No matter how well you plan, occasionally you will experience a shortage of worms due to the weather or problems with your worm bed. If you temporarily do not have worms available, you may lose some customers forever because they will cross you off their list and buy from another grower. To prevent that from happening, establish relationships with other worm growers so you can help one another to fill purchase orders. Numerous earthworm farmers have formed alliances with multiple worm growers to ensure that they always have a consistent supply for shipping. If you decide to make agreements with other growers, be sure to establish and agree on quality standards with those growers so that your company furnishes a consistent quality product no matter where it is shipped from.

Vermicast: Quality, Stability, and Maturity of Product

For most growers, making high-quality vermicast is the more important goal for their business's success. A variety of factors will affect the quality, stability, and maturity of your vermicast end product:

- Earthworm species used
- Average worm population density in the bin or pile
- Feedstocks and bedding used
- Type of processing system employed
- Length of processing time
- Duration and method of storage of finished vermicast

Pre-composted feedstock affects the available nutrient content and absence of pathogens and seeds in the vermicast. Applications of lime, sulfur, or other inorganic materials during the vermicomposting process will affect the chemical characteristics of the finished vermicast. If you vermicompost correctly and leave vermicast in the bed long enough for it to be thoroughly processed by the earthworms, the product should be stable.

Testing Vermicast

How can you determine whether you've produced good-quality vermicast? By having a sample analyzed at a laboratory, for a start. If you plan to sell your vermicast, your state may require that you obtain a laboratory analysis. Even if it is not mandatory, it is advantageous to test your vermicast so you know the quality of your product and can make process adjustments to improve it. Also, customers will want to see your product test results to help them determine whether your product will meet their needs.

The US Composting Council has developed standards for thermophilic compost. That may seem like a simple statement, but it took over 10 years for people in the composting industry (producers, users, scientists, marketers, and regulators) to volunteer countless hours haggling over and finally agreeing to a set of compost quality standards. I'm grateful for their dedicated efforts.

Unfortunately, the vermicomposting industry has no trade association or governing body to develop standards for vermicast, so there simply aren't any. But in this chapter, I share recommendations for various aspects of vermicast that have been analyzed by laboratories.

Since vermicast is not part of the composting standards, vermicast manufacturers cannot participate in a testing program developed by the US Composting Council. But I think it is important for you to be aware of it so I'm going to describe it below.

The Seal of Testing Assurance (STA) Program was created by the US Composting Council in 2000 upon the consensus of compost research scientists. They developed a compost testing, labeling, and information disclosure program that compost producers and purchasers can use to identify which compost product is most suitable for their needs. You can look for the program logo and the words US COMPOSTING COUNCIL SEAL OF TESTING ASSURANCE on a bag of compost. The logo is assurance that the producer has joined the program and has their product tested monthly or quarterly, based on production tonnage. Lab analyses are conducted by laboratories that have certified they will perform the test methods outlined in the STA Program's Test Methods for the Examination of Composting and Compost (TMECC) manual. Customers can request the Compost Technical Data Sheet that is provided with the product; it is usually kept behind the counter at the store, and if it's not there, you can contact the producer and ask for a copy. The data sheet provides the following information: feedstocks used to make the compost, soluble salts (conductivity), pH, nutrient content, organic matter content (percentage of dry weight), particle size, moisture percent, maturity (bioassay), stability and maturity indicators, inerts, select pathogens, trace metals, and directions for product use. The program's slogan is, "If it isn't STA Certified Compost, *what is it?*" You can learn more about the STA program through the Composting Council website at www.compostingcouncil.org.

Because there's no organized program for vermicast testing, you'll need to investigate your options for laboratories that will test your product. Can you get it tested locally or will you have to ship it somewhere else in your state or even out of state? Will your state department of agriculture or land-grant university do a laboratory analysis? Or will you need to use a private lab that will likely be more expensive?

In North Carolina the department of agriculture conducts high-quality testing for a low fee. They analyze your sample and provide your results online. The turnaround time is quick: Vermicast is usually analyzed within 7 to 10 days. The Agronomic Services Division conducts the following analyses: nematode assay, plant tissue analysis, soil testing, soil-less media analysis, solution analysis, and waste/compost analysis. When you send in vermicast to be analyzed, indicate clearly that you want a waste/compost analysis. If you accidentally indicate that your sample is for soil testing, the lab will use different testing techniques that will skew your results.

The division has been receiving an increasing number of vermicast samples during the past several years, so they recently added a vermicompost category to their sample form (but this may not be an option with the lab you work with). Standard analysis includes nitrogen, phosphorus, potassium, calcium, magnesium, sulfur, iron, zinc, manganese, copper, boron, sodium, aluminum, electrical conductivity (EC), carbon, pH, and carbon-to-nitrogen ratio. Other tests are available for an additional low fee. The NC Department of Agriculture (NCDA) also tests out-of-state samples (for a slightly higher fee), so you may want to consider using their lab. At a presentation I gave in the Midwest about 15 years ago, audience members begged me to take their samples with me back to North Carolina for testing, because their state DOA laboratory charged $100 per sample. At that time, NCDA didn't provide out-of-state analyses, but they do now. For information about how to test samples through NCDA and links to other lists of laboratories, see the resources section at the back of this book.

Submit to a laboratory 1 to 2 quarts (liters) of a representative sample of vermicast in a sealed plastic bag. Also called a composite sample, it must represent the entire pile of material to be analyzed. Keep the piles (or containers) of vermicast from different bins separate and test them independently because they may vary. To obtain a composite sample of a harvested pile, collect vermicast from several locations and depths within the pile. The size and shape of the pile of vermicast will dictate the number of sampling points. As a general rule of thumb, collect vermicast from at least five locations in the pile and from three depths at each location. Use a clean shovel to remove slices of material from uniform depths and deposit them into a clean 5-gallon (19 L) bucket. When you have collected all of the samples of vermicast in the bucket, mix them together thoroughly either in the bucket or on a clean tarp. Then put 1 or 2 quarts (depending on the tests you are requesting) of the composite sample into a plastic bag that can be zipped shut. Label the bag to identify the sample, and fill out the vermicast sample analysis submission form. Mail or deliver the sample to the laboratory as soon as possible. Check with the lab in advance to verify how they want the samples to be collected and see if they want the composite sample to be shipped in a container with ice packs on top of the bag.

Target Values for Vermicast Lab Analysis

The North Carolina Department of Agriculture (NCDA) recommends the target values for vermicast in lab analyses, as shown in table 9.1.

NCDA says the bottom line when reading your vermicast lab analyses is to pay attention to pH and C:N (carbon-to-nitrogen ratio). For vermicast, the limiting factors to optimal plant growth are salts, liming equivalencies, copper, and zinc.

PATHOGENS. Insignificant amounts of fecal coliform bacteria are present everywhere in the environment. Usually the levels are so low that they are not harmful to people, plants, or pets. In compost samples, fecal coliform and salmonella are tested to establish whether pathogens have been diminished to levels that are safe for humans and plants. To be considered safe, fecal coliform must be less than 1,000 MPN (Most Probable Number) per gram, and salmonella should be below 3 MPN per 4 grams.

TABLE 9.1. Target Values for Vermicast Lab Analysis

Analysis	Target Value	Notes
pH	Aim for 6.5	
EC	150–350	Values above 650 will burn plants.
C:N ratio	Not above 30:1	The 10:1 to 20:1 range is good.
Carbon	300,000 ppm or 30 percent	Material that is greater than 30 percent carbon indicates that it is not finished or stable.
Sodium	<750 ppm	
Boron	<20 ppm	Plants are very sensitive to boron.
Zinc	<300	
Copper	<50	
Iron	<15,000 ppm	Iron is rarely detrimental or toxic to plants, so it can have a broad range.
Manganese	<700 ppm	
Calcium	1–3 percent	
Magnesium	0.2–0.8 percent	
Phosphorus	0.15–1.5 percent	
Sulfur	0.1–1.0 percent	Not toxic.
Nitrogen	1–1.5 percent	This should be your target rate. A reasonable rate (acceptable) is 0.75–3 percent. If vermicast has a lot of nitrogen in it, such as 3–5 percent, it indicates that the vermicast is not finished.
Potassium	0.4–2 percent	

Source: North Carolina Department of Agriculture.

SOLUBLE SALTS. Soluble salts impact moisture availability. As soluble salt levels build up in soil, it becomes harder for plants to obtain water from the soil. An overabundance of salts can suppress seed germination, diminish yields, and impact root health, among other things. A value of less than 4 decisiemens per meter (dS/m) is generally desirable—but bear in mind that plants differ broadly in their salt tolerance.

Storing Vermicast

Store vermicast at a 40 to 50 percent moisture level (by weight). If the moisture content of the vermicast is below 35 percent, it becomes dusty and hard to handle or hardens to a bricklike consistency. If the moisture level exceeds 60 percent, it clumps together.

For storage, place vermicast inside a bag or a bin with a lid. The container needs to allow the passage of air to keep the microorganisms in the vermicast alive, yet

FIGURE 9.9. These hand-built plastic tubes are used as bagging sleeves at The Worm Farm in California. A worker inserts a sleeve into a bag and shovels vermicast to fill the bag with 1 cubic foot (0.03 m³) of vermicast.

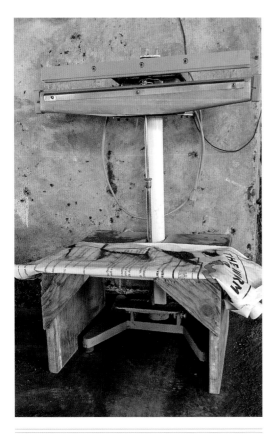

FIGURE 9.10. This device seals bags of vermicast at The Worm Farm in California.

too much aeration will facilitate moisture evaporation; the material will dry out, which will kill the microbes. If you prick tiny holes in the container, they will allow the passage of air yet prevent water vapor molecules from escaping.

Store the vermicast away from direct sunlight, in a cool, dry location. Vermicast may be stored for approximately 18 to 24 months.

Packaging Vermicast

The moisture content in harvested vermicast will vary over time, and that changes the weight of the product. Therefore, many worm growers sell by the cubic foot or yard, not by weight.

There are pros and cons of selling in bulk and selling in bags. Some worm farmers choose not to sell in bags because it would cost them more than selling in bulk. Some farmers sell both in bags and bulk. Bagging machines can be quite pricey, but I've seen some ingenious, inexpensive methods of bagging at the worm farms I've visited. For example, at The Worm Farm in California, they simply cut a sturdy plastic tube into a size that will hold 1 cubic foot (0.03 m³) of vermicast. They fasten two knobs across from each other at one end so they can easily maneuver the tube, as shown in figure 9.9. They insert the tube into a 1-cubic-foot bag, and the tube holds the bag open so they can fill it.

There's no one type of bag that is best for vermicast. One worm farmer uses a poly-lined paper bag from the pancake industry to bag his vermicast.

Another uses poly-lined paper ice cream pint containers. Two bag manufacturers that worm farmers commonly order from are Trinity Plastics Inc. and Salerno Plastic Films and Bags.

To seal bags, some people use staplers or zip ties. Many use heat sealers like the one shown in figure 9.10.

Product Labeling

Every state in the United States and the territory of Puerto Rico has its own fertilizer regulatory program administered through the state's department of agriculture (DOA). The Association of American Plant Food Control Officials (AAPFCO) is the technical recommending body for all of the state departments of agriculture, Puerto Rico, and Canada. The AAPFCO works to ensure uniform definitions, rulings, and enforcement practices regarding the distribution of fertilizers, soil amendments, and lime. This includes sampling and analytical methods for fertilizer, inspection techniques and enforcement practices, promotion, adequate labeling, and safe use of fertilizers. All of this is done in an effort to promote the safe and effective use of fertilizers, and to protect soil and water resources.

Currently, 38 states have laws regulating the sale and distribution of soil amendments. *Soil amendments* are delineated as products that are mixed into topsoil to promote healthy plant growth.

Every state has fertilizer laws except Alaska and Hawaii. It is against the law to make false or misleading statements for products that fall under the category of fertilizer. This includes deceptive trademarks or brand names. If you are found in violation, the state DOA can suspend, revoke, or terminate your license or registration for that fertilizer product.

Although vermicast is rarely mentioned in state regulations, most officials who are regulating fertilizer and soil amendment products put it in the category of compost. They understand that vermicast and compost are created by diverse methods and the products have differences; however, because rules for vermicast don't exist, state officials apply compost regulations to them. There are also regulations regarding manure. Unmanipulated manure is not regulated if you give it away. However, if you are selling manure, it is considered fertilizer. And compost or vermicast that is manure based is considered manipulated manure and thus is deemed a fertilizer.

Fertilizers are basically defined as products that are able to supply plant nutrients. The primary plant nutrients are nitrogen (N), phosphorous (P) and potassium (K). As you probably have seen, bags of fertilizers are labeled with a three-number analysis equivalent to N, P, and K. The numbers indicate the percentage of the net weight that is made up of these three nutrients. For example, a 50-pound (23 kg) bag of fertilizer labeled 10–10–10 contains the equivalent of 5 pounds (2.3 kg) of nitrogen, 5 pounds of phosphate, and 5 pounds of potash. (This is calculated by multiplying 50 by 0.10.) Phosphate and potash are available plant-available forms of phosphorous and potassium, respectively.

If you include nutrient levels on your product labels, you have made a fertilizer claim that is subject to regulation. You must comply with the labeling regulations of each state in which you plan to sell your product. The more states you want to sell your product in, the harder it becomes to label your package because the specifics of the regulations vary from state to state.

Let's say that you decided to include an NPK listing of 1.0–0.5–0.5 on your labels for your bagged vermicast. As far as the state DOA is concerned, you have declared that your product is a fertilizer containing those levels of nutrients. A department of agriculture official can randomly pull your product from the store shelf and test it. If your product is analyzed by the state DOA and it falls short of the guaranteed analysis in any ingredient (say, your nitrogen tests as 0.8 instead of 1.0), a penalty will be assessed in accordance with the state's regulatory provisions. You can be slapped with a stop-sale order, and it can take months and a lot of money and effort to get your product back on store shelves.

There is an annual tonnage fee and registration fee to register products. If you wish to make a claim that your product is a fertilizer, then the product label must list the net weight (not volume) of the product. Unless your product is registered as a fertilizer, it is unlawful to provide any nutrient data about it. This includes laboratory analyses.

For registration purposes, most states consider vermicast as a Specialty Fertilizer. This means that the product is distributed chiefly for nonfarm use (golf courses, home gardens and lawns, nurseries, greenhouses, municipal parks, and so on).

If you register your product as a fertilizer, you can also make soil amendment claims. However, if you are following the guidelines for your product to be considered a soil amendment, you cannot make fertilizer claims.

You cannot make the following claims on any vermicast product label: growth hormone, disease suppression, pest suppression, or microbial claims. If you want to make beneficial microorganism claims, your product must be registered as a soil amendment and state the names of the species present and their numbers (it's basically impossible to do this accurately, as what you see under a microscope at a certain time may not match what you see at any other time).

It's unlawful to claim that your product provides disease resistance for plants. You can provide a published article about disease resistance along with your product, but you cannot print a disease-resistance claim on your bag label or post it on your website because regulators will interpret that you have declared your product to be a "plant pesticide."

Twice yearly (subsequent to their midyear and annual meetings), the Association of American Plant Food Control Officials publishes *The AAPFCO Insider*. In volume 4, published in 2012, AAPFCO announced that certain beneficial label claims for vermicompost were moved to official status (these claims can now be included on a product label):

- Improves soil structure and porosity—creating a better plant root environment
- Increases moisture infiltration and permeability and reduces bulk density of heavy soils—improving moisture infiltration rates and reducing erosion and runoff
- Improves the moisture-holding capacity of light soils—reducing water loss and nutrient leaching and improving moisture retention
- Improves the cation exchange capacity (CEC) of soils.
- Supplies organic matter
- Aids the proliferation of soil microbes
- Supplies beneficial microorganisms to soils and growing media (only for products providing minimum microbe content guarantees)
- Encourages vigorous root growth
- Allows plants to more effectively utilize nutrients
- Enables soils to retain nutrients longer
- Contains humus—assisting in soil aggregation and making nutrients more available for plant uptake
- Buffers soil pH
- Is a source of macro and micro nutrients (for fertilizer-registered products, not soil amendment–registered products)

Tea or Aqueous Extracts

Vermicast tea is made by using water to extract microorganisms, organic matter, and nutrients from vermicast. The intention is to transfer these components into an aqueous solution and sustain or increase the live beneficial microbes. If you are interested in selling vermicast tea, you could start by reading the profiles about worm farmers who are selling it.

It is vital to use the highest-quality vermicast or compost, or both, to make tea. If the solid material is inferior, it will make deficient tea. Whatever problems may exist in the vermicast or compost (pathogens, high salt levels, an abundance of anaerobic microorganisms) will be increased in the tea. The microbiology of the tea will be influenced by the original microorganisms, nutrient content, water quality, availability of oxygen, duration of brewing, water temperature, and supplemental nutrients.

There are two primary methods of extraction. Non-aerated compost tea (NCT) is a passive approach that has been in use for centuries. It involves steeping the compost or vermicast in water for 14 days and occasionally stirring it if desired. Aerated compost tea (ACT) involves using an aerator to oxygenate the concoction for 12 hours to 3 days. Frequently people enhance the mixture with nutrient additives or fermentation products with high microbial content, such as fish emulsion, sea kelp, rock dusts, oatmeal or bran, or fish hydrolysate. Along with many others, I warn people to not add sugar

(white sugar, molasses, corn syrup, cane sugar, or maple syrup), because although it will increase the microbial population in the tea, it will also boost the growth of pathogenic microorganisms such as *E. coli* and *Salmonella*. Researchers note that adding sugar to a tea brewer causes a chain reaction of a huge spike in the microbial population, which uses up the oxygen, causes the microorganisms to die, and creates toxicity.

Because chlorine kills beneficial microorganisms, you should use nonchlorinated water to make tea. Well water and rainwater are obviously nonchlorinated options. If you only have chlorinated water available, you can fill the tea maker to the desired level and let the chlorine off-gas (evaporate) for 24 to 48 hours. For optimal results, maintain the water temperature at 70 to 75°F (21–24°C) while the tea is brewing, as solubility of nutrients is dependent on the temperature of the water.

FIGURE 9.11. This 25-gallon (95 L) tea brewer by Growing Solutions will produce enough vermicast tea for 1 to 5 acres (0.4–2.0 ha). If you place the brewer on a pallet, you can use a forklift to lift it high enough to fill up 30-gallon (115 L) drums in customers' trucks. The orange hose is used to fill the drums as well as gallon jugs.

Many worm farmers make their own tea brewers. The simplest method is to fill a clean 5-gallon (19 L) bucket with water, put vermicast in a fine-mesh bag, and submerge the bag in the water. You can use an aquarium pump to aerate the tea.

You can also purchase commercially manufactured equipment with the capacity to brew as little as 5 gallons or up to 500 gallons (19–1,895 L) of tea. The manufactured tea brewers use a variety of methods to aerate the tea, such as stirring, recirculating, forced air, or extractors.

After every batch of tea is made, it is imperative to rinse out the container and then scrub it with a brush using a hydrogen peroxide solution to sanitize it. If you don't clean the receptacle, bio-slime will form, which is full of anaerobic microbes that will end up in your next batch of tea. This can harm or kill plants! If bio-slime dries in your tea brewer, the only way to clean it is to scrape off the film using a single-edge razor blade. Keep this in mind when constructing or purchasing a tea brewer. If you will not easily be able to clean every nook and cranny of the device, do not buy or make it!

Vermicast Tea Versus Leachate

Leachate is *not* the same thing as vermicast tea. Leachate is the dark liquid that leaks out of the bottom of some worm bins. Leachate has passed through undigested, anaerobic areas of the worm bin, so it could contain several undesirable components. These include sulfides, acids, high salts, pathogens, or anaerobic microorganisms that could be harmful to people, animals, or plants. I call it the Stinky Mystery Liquid because it is usually smelly and you do not know what is in it. Leachate should never be used on food crops and is not recommended for indoor or sensitive plants. Why take a risk with it?

If you are operating your vermicomposting system properly, there will not be any leachate to worry about. If excess liquid is coming out of the bottom of your worm bin, it's a sign of incorrect management, such as pouring water into the bin, adding too much food, not providing enough carbonaceous bedding, or putting in too many foods that contain excessive moisture (such as canned or cooked food or coffee grounds). If your bin does leak leachate, remove it and pour it on the soil outdoors around vegetation that does not have value, such as a weedy area. Do not pour leachate into a storm drain, stream, pond, or other body of water.

Using Tea

Vermicast tea affects plants faster than solid vermicast that is mixed into soil. Tea can be applied as a soil drench or a foliar application (sprayed on plant leaves).

Tea is increasingly being used by vineyards, golf courses, organic farms, greenhouses, lawn maintenance companies, community gardens, and parks as a method of integrated pest management (IPM). They use vermicast tea at concentrations of 5 to 20 percent. Two gallons (7.6 L) of vermicast added to 10 gallons (38 L) of water yields 20 percent vermicast tea.

Some research studies have shown that applying vermicast tea may suppress plant diseases and insect pests and increase the vitality of plants and soils. Vermicast tea has been effective in enhancing plant growth in concentrations of less than 50 percent. Research by Dr. Norman Arancon revealed early seedling germination if seeds were soaked in vermicast tea for one hour and then planted immediately afterward. You may think, *I bet the seeds would germinate even quicker if the seeds were soaked longer,* but Dr. Arancon already tested that. He soaked separate seeds in vermicast tea for 1 hour, 4 hours, 8 hours, 12 hours, and 24 hours. He staggered the starting times of the soaking so that he was then able to plant all of the seeds at one time. The best results were from the seeds that had soaked in tea for one hour.

Scientific studies of the efficacy of teas show mixed results. This may be due to several factors, including the quality of the vermicast, how the tea was prepared, how long it was stored before use, and how the tea was applied. And as with solid vermicast, distinct plant species will respond differently to extracts from diverse sources.

Each batch of tea can be different as well. Factors affecting tea content include differences in the vermicast being used, the length of brewing time, vermicast-to-water ratio, aeration differences, and alterations in supplemental materials. If the equipment wasn't cleaned thoroughly between brewing, it will also affect the subsequent batch.

For the best results, vermicast tea should be used immediately after it is brewed and definitely within four hours of brewing.

Here is another interesting bit of information about vermicast tea: Dr. Arancon tested the microbial activity in vermicast extracts in cold storage. He aerated vermicast tea for 24 hours and then put it in a refrigerator set at 39°F (4°C) for 21 days. The microbial population reached the highest amount at 14 days, and it plummeted at 21 days.

The following are recommended spray times for vermicast tea applications.

- Every one to two weeks
- Early in the day before it gets hot
- On a cloudy day to prevent microbial loss from ultraviolet radiation
- After irrigation or rain

For foliar application, be sure to use a filtering system to prevent the sprayer from getting clogged. Spray the surfaces of leaves thoroughly to ensure you inhibit plant pathogen colonization.

Making and using vermicast tea is a relatively new endeavor, and more scientific research is required to understand how it can be the most effective. If you are experimenting with vermicast tea, continue to gather information about the latest published research results.

CHAPTER 10

Vermicomposting Operations Around the World

Now that you've learned about the nuts and bolts of planning, starting, and managing a vermicomposting operation, let's take a tour of vermi-operations around the world. In these profiles of vermicomposting setups at businesses, schools, institutions, and a variety of mid- to large-scale commercial worm farms, I feature the unique methods each uses to produce vermicast and other products.

Many people do not realize that vermicomposting is taking place worldwide. As I mentioned in the introduction to this book, folks in more than 100 countries have contacted me by email to ask for advice about vermicomposting. They live on every continent except Antarctica (go figure). Their questions are similar to the ones I receive from people throughout the United States. A couple of dozen emails annually are from people who have earned their doctorate degrees conducting vermicomposting research and want me to hire them. They think I have a fancy laboratory staffed with postdocs, but sadly that is not the case. I do love hearing from everyone, and I am constantly amazed at and impressed with the innovative vermicomposting systems that people are utilizing across the planet.

As you read along, I hope you will get ideas for methods or equipment you want to try in your existing or future vermicomposting operation. Enjoy!

Bittercreek Alehouse & Red Feather Lounge

Boise, Idaho

Even though unused space is hard to find at a restaurant, this alehouse figured out how to set up a vermicomposting operation using space in an alley and the restaurant's basement. In 2006 the Bittercreek Alehouse & Red Feather Lounge bar/restaurant in downtown Boise began vermicomposting their organic waste in the basement. They bought two 14-by-4-foot (4.3 × 1.2 m) continuous flow-through bins that cost $12,000 apiece. The floors of the bins were metal grates with 4-by-2-inch (10 × 5 cm) holes. The bins were set up with 6 inches (15 cm) of shredded cardboard and paper bedding, and 100 pounds (45 kg) of earthworms. Every week the restaurant/bar produces 250 pounds (115 kg) of food scraps and 40 to 50 pounds (18–23 kg) of coffee grounds mixed with shredded paper menus and beer lists. These organic materials are collected in designated buckets in the kitchen and then pre-composted for two weeks in two thermophilic compost bins in the alley behind the restaurant. The compost bins are Suncast Tumbling Composters, each of which has a footprint of 6 feet by 4 feet (1.8 × 1.2 m) and a capacity of 6.5 cubic yards (5 m³) or 200 pounds (91 kg). The restaurant chose these bins because regulations required that they have enclosed containers for composting. Once a week they apply a 1-inch (2.5 cm) layer of pre-compost to the top of each worm bin. Vermicast is harvested from the bottom of the bins, where it falls to the floor and is swept up with a broom. They use a trommel screener to screen the vermicast, which they sell or trade to local farmers who supply their produce. The restaurant developed a popular additional source of income based on their vermicomposting activities by selling earthworms, vermicast, worm bins, T-shirts, onesies for babies, tote bags, books, worm bin bedding, and thermometers online and in-house. They also provide tours of their vermicomposting operation.

The Len Foote Hike Inn

Dawsonville, Georgia

The Len Foote Hike Inn is a unique state park facility that is accessible only by foot. The Leadership in Energy and Environmental Design (LEED) inn was built in 1998 with sustainability in mind and featured vermicomposting, solar panels, composting toilets, and rainwater harvesting. Food preparation scraps and plate scrapings from meals served to guests were fed to *Eisenia fetida* earthworms. In addition to food residuals, the inn vermicomposted shredded cardboard and office paper, discarded cotton and wool clothing, and even cotton mopheads. There were 10 large earthworm bins in the basement that processed more than 1,500 pounds (680 kg) of organic materials per year. Staff members harvested vermicast regularly from the bins and incorporated it into the inn's organic garden and soil around native plants. Inn staff also taught guests how to set up their own worm bins.

The Bishop's Ranch

HEALDSBURG, CALIFORNIA

This retreat and conference center has a vermicomposting system to recycle food waste generated by the conference center kitchen. Their previous conference center manager Rick Kaye set up the vermi-system in 2009.

When I visited in 2017, the ranch had six worm bins made out of sections of plastic culvert pipe. Each bin is fed 20 pounds (9 kg) of food waste once a week, in all about 3 tons (2,720 kg) annually. About 1.5 cubic yards (1 m³) of vermicast is harvested from the worm bins three times a year.

FIGURE 10.1. Plastic culvert pipes are cut to be used as worm bins to turn food wastes into vermicast for the gardens at The Bishop's Ranch in California.

Evergreen State College and Cedar Creek Corrections Center

Olympia, Washington

Here's an example of vermicomposting at two institutions in northwestern Washington State: a college and a prison.

The Evergreen State College began vermicomposting at their organic farm in the early 1990s. Initially, they vermicomposted food waste from dining services and a local restaurant in hand-built wooden bins a total of 50 feet (15 m) in length. Beginning in 2008 Evergreen changed their campus organics management to two streams sent to separate locations. Most campus food waste from dining halls and student housing, including compostable dinnerware, was shipped to Silver Springs Organics, a commercial composting facility located a short distance away in Rainier, Washington. Approximately 480 gallons (1,815 L) of food residuals was collected each week from dining services and taken to Evergreen's organic farm composting facility. First it was composted in an aerated, in-vessel thermophilic composting unit and then finished in a vermicomposting system. They had four 8-foot (2.4 m) modular continuous flow-through vermicomposting bins that were connected and a total of 32 feet long by 5 feet wide (9.8 × 1.5 m). Up to 125 gallons (475 L) of compost was applied to the top of the worm beds weekly. Vermicast was released through the mesh bottom by breaker bars into trays below and used on the organic farm and community gardens.

Cedar Creek Corrections Center (CCCC) partnered with the Evergreen State College and the Department of Corrections to create the Sustainability in Prisons Project. Among its multifaceted sustainability initiatives, prison administrators decided to develop a vermicomposting program that would divert food waste from disposal, improve regional groundwater quality, and produce organic fertilizers for the prison's vegetable garden. The 400-inmate facility began vermicomposting in 2004 by building a stacking-shelf vermicomposting system using recycled lumber. Six shelves held three stacked earthworm bins apiece. Each shelf was 3 feet long by 3 feet wide by 7.9 inches high (91 × 91 × 20 cm). Earthworms were fed approximately 3,000 pounds (1,360 kg) of food residuals per month from the inmate kitchen, staff food scraps, coffee grounds, and garden waste. Vermicast was harvested every two or three months and used in the facility's gardens to grow vegetables that supplemented prisoners' daily meals. Prisoners designed and built a hand-cranked sifting machine to separate the earthworms from the vermicast. It cost an average of $98 per ton to haul and dispose of solid waste, so vermicomposting at CCCC saved almost $2,000 per year. By vermicomposting instead of disposing of food down the drain, CCCC reduced its per-capita water consumption by about 25 percent.

Southern Illinois University

Carbondale, Illinois

Southern Illinois University began vermicomposting in 2006 using a $150,000 grant from the Illinois Department of Commerce and Economic Opportunity to cover their start-up costs. The operation was a joint effort between the Plant and Service Operations and the College of Agricultural Sciences to research and recycle university food waste. The university had a setup similar to Evergreen College's vermi-system, with four 8-foot (2.4 m) modular continuous flow-through bins that were connected to create a system 32 feet long by 5 feet wide (9.8 × 1.5 m). The bins were housed in a roughly 5,000-square-foot (465 m²) pole barn 2 miles (3.2 km) from the main campus. About 12,800 pounds (5,805 kg) of vegetable waste was collected each year from a campus dining hall and put through a pulper to grind the food and extract liquids.

Medical University of South Carolina

Charleston, South Carolina

In the late 1990s the recycling coordinator at The Medical University of South Carolina (MUSC) decided to increase their waste diversion rate by implementing a food residuals recycling program. Constrained by limited space in its urban location, the university opted for vermicomposting. In 1999 MUSC obtained funding from South Carolina's Department of Health and Environmental Control and that department's Energy Office, in addition to the Sustainable Universities Initiative. They used $25,000 to purchase a flow-through vermicomposting reactor, an organic materials shredder/mixer, and earthworms. Another $31,000 was spent to house the reactor in a 24-by-18-foot (7.3 × 5.5 m) ventilated building with a sloping acrylic floor that could be easily cleaned. The vermicomposting system measured 18 feet long by 7 feet wide by 5 feet high (5.5 × 2.1 × 1.5 m). Inside the system were two 8-foot (2.4 m) modular beds and a 2-foot (61 cm) section for hydraulic equipment and an air conditioner. Cafeteria kitchen staff placed pre-consumer food waste in a wheeled 45-gallon (170 L) container. A recycling team member picked up the container daily and took it to the vermicomposting building for weighing and processing. The weight varied according to the types of food prepared, but on average they collected approximately 115 pounds (52 kg) per day. The food waste was emptied into the organics shredder/mixer, and cardboard was ground up with the food to achieve the proper moisture level (originally they used newspaper, but they discovered it got too soggy). The mixture was loaded into the flow-through reactor using a conveyor belt attached to the shredder. It took one hour daily to collect the food waste, weigh and shred it, and feed it to the earthworms.

Another 30 minutes per week was spent inspecting the reactor environment and the vitality of the earthworms. They monitored the bed temperature daily and found that 72°F (22°C) was ideal. Vermicast was harvested from the reactor every two to four weeks using a hydraulically operated breaker bar to release it to the floor below. A broom and squeegee were used to collect the vermicast, which totaled about 500 pounds (230 kg) per month. MUSC's grounds department used the vermicast as a soil conditioner, reducing their need for commercial inorganic fertilizers.

Broad River Correctional Institution

Columbia, South Carolina

In 1998 there were 1,000 inmates at Broad River Correctional Institution. The prison had two greenhouses and beehives, and they raised plants for the state nursery association for semiannual identification tests. They began vermicomposting that year and invited me to visit and do some troubleshooting. The prison paid a company $2,000 for a worm bin, installation, and 120 pounds (54 kg) of earthworms. The bin was manufactured with wood and insulated panels on the inside. It measured 34 feet long by 7 feet wide and 20 inches deep (10.4 × 2.1 × 0.5 m). The bin had a divider running down the center so each side was 3½ feet (1 m) wide, which made it easier to hand-feed and harvest the bin. The bin was set up in the open prison yard, and the inmates installed a screen on the bottom to keep out moles. They also added greenhouse mesh suspended by metal poles to shade the worms from the intense sun. Food waste and yard waste were collected, and the yard waste was shredded using a rotary tiller. The mixed wastes were pre-composted, and when the temperature of the compost pile dropped below 80°F (27°C), about 4 inches (10 cm) of compost was applied to the top of the worm bed at each feeding. The vermicast was used for landscaping.

Wright-Patterson Air Force Base

Dayton, Ohio

Vermicast is a great natural fertilizer for golf courses. Wright-Patterson Air Force Base began vermicomposting in July 2002. They vermicomposted a daily average of 500 pounds (230 kg) of vegetable food residuals and newspaper, which saved them $25 a day on hauling and waste disposal fees. Approximately 300,000 earthworms devoured the feedstocks in a climate-controlled flow-through worm bin. The bin and 250,000 earthworms were acquired at no cost from another military base that could no longer use them. Measuring 16 feet long by 8 feet wide by 5 feet tall (4.9 × 2.4 × 1.5 m), the worm bin included air-conditioning and heating capabilities that enabled it to maintain a constant temperature. The digester was housed indoors to protect it from the elements and fluctuating temperatures. Recycling center staff discovered that the earthworms thrived when the system operated at 70°F (21°C). The labor requirement was one hour per day for one person to operate the system and care for the earthworms. The vermicast was used on the base's golf course and grounds to save money on soil amendments and reduce fertilizer runoff into ground- and surface waters.

FIGURE 10.2. Madrone Elementary School in Santa Rosa, California, has five 2-foot-tall (61 cm) culvert worm bins. The bins have custom-made wooden lids covered with elasticized shade cover cloths.

Worm Wizards of Waste

CALIFORNIA

The Worm Wizards of Waste grew out of the Compost Club in Sonoma County, California (described in chapter 1). The group is part of the North Coast Resource Conservation & Development Council. In 2017 there were 35 school-wide vermicomposting systems in Mendocino, Lake, Sonoma, Napa, and Marin Counties in California.

Volunteers from the Worm Wizards visit schools to help analyze their cafeteria waste situation and then develop a low-tech vermiculture system. For most of the schools, they created worm bins made from plastic culvert pipes (like the ones Rick Kaye created at The Bishop's Ranch). The corrugated culvert pipes are smooth on the inside and have thick bulges on the outside that help provide insulation, keeping the interior cooler during the summer and warmer in the winter. The original culvert bins were set on frames made from ¾-inch (19 mm) CDX plywood with 15- to 18-inch (38–46 cm) legs to make them rodent-proof. In 2017 the groups switched to building tables and setting the culverts on top. Healdsburg High School students learned construction skills by building 20 tables. Two bins were set on top of each table. Newer culvert bins had three or four corrugations instead of taller pipes with five or six bulges. The pipes cost $50 to $60 per foot.

They made round wooden lids for the bins and elasticized shade cover cloths to hold the wooden lids in place. The lids and shade cloth served to keep out flies and other critters. They discovered at first that raccoons were able to open the elasticized cover cloths, so they attached plastic buckles to tighten the covers and foil the animals. The Bishop's Ranch donated two 5-gallon (19 L) buckets of worms and organic materials to start new bins at schools. Moving the worms along with their bedding reduced the shock to the worms of being relocated.

When club volunteers began working with a school, they calculated how much food waste was generated per year. They introduced the program to the students with a lesson called Recycling Changes Everything. They used three types of lunches in the demonstration—Lunchables (a packaged lunch manufactured by Kraft Foods), Laptop Lunch (a reusable bento lunch box), and a paper bag lunch—and they sorted the recyclables, organic materials, and landfill items. The team identified a lead classroom to take responsibility for the lunch food waste recycling program. Then the group obtained grants from community groups, Rotary Club, University of California Cooperative Extension, county waste management association, and private donors.

It is common for schoolchildren in California to eat lunch outdoors on large picnic tables. After lunch they sort their recyclables and organic materials. At

How to Start a School Vermicomposting Program

Here's the Compost Club's advice on the steps to take to initiate a vermicomposting program:

- Assemble a school team.
- Calculate food scrap generation.
- Estimate the number of worm bins needed.
- Raise funds to pay for equipment and worms.
- Educate users (students, staff, custodians).
- Train a lead person.
- Hold a community build day to install the worm bins.
- Help schools sell their vermicompost at farmers markets or events.

Madrone Elementary School the students used to have lunch first and then playground times but because they were in a hurry to play, the students didn't sort their lunch waste very well. So the school reversed the order; now the students have playground time before lunch.

Salmon Creek School

FREESTONE VALLEY, CALIFORNIA

Salmon Creek School is on 80 acres (32 ha) in a redwood forests, and it has extensive vegetable, fruit, and pollinator gardens, and a straw-bale greenhouse. Most of the food served at the school is grown onsite or at local farms. When I visited in 2017, they had an open-air shed for vermicomposting that had a roof and shade cloth sides. The shed was made from trees grown onsite that were milled. Salmon Creek had four wooden worm bins that were constructed by volunteers. Each bin had four lids, creating four sections with a screen crossing the middle of the bin to use for horizontal harvesting. The lids were held open with hooks attached to chains hanging from the roof. Two sections of one bin were fed lunch waste each day. Food waste was placed on top of the organic material in the bin and then covered with paper egg crates as shown in figure 10.4.

FIGURE 10.3. The worm bins at Salmon Creek Elementary School are both durable and attractive.

FIGURE 10.4. The screen divider inside the worm bin allows for a horizontal harvesting system.

Encinitas Union School District

Encinitas and Carlsbad, California

SCRAP stands for Separate, Compost, Reduce and Protect. Encinitas Union School District (EUSD) board of trustees approved of the SCRAP waste diversion program for all nine of its schools. The SCRAP program required all Encinitas schools to implement full-stream recycling and food scrap composting. Customized lunchtime waste sorting units (called SCRAP Carts) on wheels were built with wood for each school. All students, staff, and custodians were trained to use the equipment. The SCRAP Carts had six labeled holes in the top: The first was labeled CITRUS ONLY, the next WORM COMPOSTING ONLY; then came REDEEMABLES for plastic bottles and aluminum cans, and MIXED RECYCLING for paper bags, cardboard, hard plastic, and tinfoil. The fifth hole was labeled MILK CARTONS, and the last was LANDFILL TRASH. Some schools generated less than two garbage bags of trash destined for the landfill at lunchtime. The SCRAP Program reduced lunchtime waste by 85 percent district-wide (from 100 garbage bags per day to 18). In 2013 EUSD's SCRAP Program received a regional award from the California Resource Recovery Association for Best Waste Prevention Program.

Jefferson County–DuBois Area Vocational-Technical School

REYNOLDSVILLE, PENNSYLVANIA

The career and technical high school known as JEFF TECH received a grant from the Pennsylvania Department of Environmental Protection and began vermicomposting in 2011. They used the grant to purchase materials for worm bins, a trommel screen harvester, and earthworms. Quiet Creek Herb Farm & School of Country Living built five worm bins, then provided the worms and training for vermicomposting. In the cafeteria food scraps were deposited into a large garbage can with a small opening. At the end of the day, Culinary Arts students weighed the food scraps and placed them in one of the five worm bins. Science students harvested vermicast from the worm bins and spread it on school gardens. In 2015 JEFF TECH vermicomposted 2,500 pounds (1,135 kg) of food residuals and 100 pounds (45 kg) of newspaper. In September 2015 they harvested 110 gallons (415 L) of vermicast.

Sierra Heights Elementary School

RENTON, WASHINGTON

Sierra Heights Elementary School began a school-wide vermicomposting program in 2007 for their 600-plus students. A fifth-grade teacher obtained an $8,000 grant from the Washington Department of Ecology and partnered with the King County Green Schools project after obtaining approval from the necessary administrators and staff. King County Green Schools staff helped train school staff about the care of earthworms, the impact of reducing waste going to the landfill, and changes to the lunchroom routine. Students were selected to collect, weigh, and deposit food waste in the worm bin. Parents played a big role in sorting lunchroom waste and managing the worm bins. They also ensured that the worms were fed and cared for during summer months when school was not in session.

The vermicomposting project was kicked off with an all-school assembly. It began with a presentation by King County Green Schools staff about how food waste affects landfills and what students can do at home and school to reduce the impacts. A group of students did an interactive activity teaching peers how to reduce, reuse, and recycle. Another set of students taught about earthworms, including their anatomy, lifestyle, and needs of worms used for vermicomposting, ending their presentation with an interactive quiz. Still another group presented a lesson on What Worms Can and Can't Eat and ended with a game-show-type challenge for the school principal to determine which foods should go in a worm bin or trash can. The last group demonstrated the changes to the lunchroom routine. They showed how, after being dismissed

School Vermicomposting Recommendations

- Develop a memorandum of understanding signed by principals, site coordinators, and district staff to clarify roles and commitments.
- Assemble a team to monitor the waste stream and maintain the worm bins.
- Begin the program on the first day of school instead of introducing it later on.
- Hold an all-school assembly to kick off the project.
- Start with younger students (who are more willing to adopt new practices).
- Don't add citrus, meat, or dairy products.
- Place worm bins in the shade.
- Create a routine monitoring and maintenance schedule for the bins.
- Ensure a steady supply of worm food and care during the summer.

from lunch, students could sort their waste into separate containers for worm feeding, recycling, liquids, and garbage. They set up a map of the lunchroom with locations for signs and the sorting containers and invited volunteers from each grade level to come on stage and try the new sorting routine. The group ended the assembly by getting everyone to sing the "Reduce, Reuse, Recycle" song that had been taught earlier in the program.

A waste audit was conducted to determine the quantities and types of lunchroom waste that was generated and establish a baseline, so they could measure their waste reduction efforts. They collected bags of lunchroom waste, weighed them, and then sorted the contents into worm chow, recyclables, liquids, and trash. The sorted materials were weighed to determine the percentage of the total waste.

After reviewing assorted designs for worm bins, they decided to buy a manufactured continuous flow-through worm bin that was 5 feet wide, 6 feet long, and 4 feet high ($1.5 \times 1.8 \times 1.2$ m). It required 110 volts of alternating current (AC) electricity and a forklift to move the bin. The bin was placed in a shady location out of direct sunlight. Even though it had two lids, rain still got through, so the school had to build an A-frame awning over the bin to keep it dry.

During each of the five daily lunch periods, four students helped their peers sort their lunch waste into separate containers. Each container was weighed to record their waste reduction efforts. By keeping track of waste, they were able to determine which school hot lunches produced more waste and cut back on them. A whiteboard was set up in the cafeteria for students to post daily announcements about how the worms were doing, concerns or information about recycling efforts, and what items from the school hot lunch could be fed to the worms.

Worm Endings Unlimited

Napa, California

Debbie Stevens turned her vermicomposting hobby into a unique business in Napa, California. Much of Debbie's sales are at the local farmers market, where she sells vermicast, *Eisenia fetida* earthworms, vermicast tea, and worm bins. Most of the items for sale have been recycled or are contained in eye-catching repurposed packaging.

Debbie vermicomposts in seven wooden wine packing boxes that are 42 inches by 48 inches by 16 inches deep (1.1 × 1.2 × 0.4 m). She starts a box with leaves, paper, and sawdust on one side as a wedge and then adds to it sideways. She feeds the worms food waste, yard waste, and paper towel or tissue rolls. To harvest, Debbie uses a scoop to get vermicast out of the box and then runs it through a homemade trommel screener.

Debbie makes vermicast tea in a Growing Solutions tea brewer using vermicast and water; she sells her tea for $10 a gallon. She washes out used gallon jugs to hold the vermicast tea. To remove labels from the jugs, she covers the paper label with creamy peanut butter. After 24 hours she can quickly scrape off the label.

FIGURE 10.5. Worm Endings' hand-built trommel is operated by a small motor.

She sells vermicast in a 1-gallon bag for $10, a 1-quart bag for $5, 1 pint for $2, and less than a cup in a snack bag for $1. All of the bags are plastic ziplocks, and Debbie uses adhesive labels with her business name, logo, and contact information. She sells a 5-gallon bucket of vermicast for $25. Of course, the buckets are reused, as well.

Her husband, Dan Grant, makes worm condos out of used wooden wine boxes by removing the wooden bottom and slots and installing screens. They are sold for $25 per box and are stackable.

Debbie also sells worm bins made from 50-gallon wine barrels. Dan cuts the barrels in half lengthwise and mounts each one on a wooden stand with a lid and lip. They sell for $206. All of the worm bins are painted inside with linseed oil to treat the wood.

Debbie buys worms from a local worm farm and resells them. She sells

FIGURE 10.6. Debbie Stevens and Dan Grant turn wine barrels and boxes into worm bins for sale. Worm Endings' repurposed-wine-barrel worm bin sells for $206.

a pound of worms for $29 in a reused ½-gallon (1.9 L) paper milk carton covered with a nylon stocking. The worms are bedded in vermicompost, paper, and sawdust. Her busy season is mid-January until the end of August. She doesn't sell anything for four months of the year and goes on vacation. I visited Debbie in August 2017, right before she and Dan headed out on another trip.

The Worm Ladies of Charlestown

RHODE ISLAND

Nancy Warner is a fiber artist who raises Angora rabbits so she can shear, spin, weave, and knit the angora hair into her art creations. In the early 1990s Nancy researched methods for managing the manure and flies that collected under the rabbits' cages. She discovered vermicomposting, and after it rapidly solved her manure management problems, she was hooked. Also an avid gardener, Nancy was impressed with the amazing effects the vermicast had on her vegetables and flowers. Two of Nancy's friends (Lois Fulton and Susie Fehrmann) became enthralled with vermicomposting as well, so the three of them decided to start a business, and that was how The Worm Ladies of Charlestown, Inc., got started. One of their first marketing ideas was to print business cards and hand them out at a flower show. Then a local newspaper wrote an article about their new business, and the phone started ringing off the hook. Their first sales were to garden clubs, and then they moved into farmers markets. They expanded their customer base into Connecticut and Massachusetts.

As their business grew, they added more products and services. They were struggling to keep up with the demand for worms and vermicast, and then marijuana growers started buying as much as the Worm Ladies could supply. Nancy and her partners began receiving orders for 100 pounds (45 kg) of worms and a cubic yard (0.8 m³) of castings, so Nancy obtained a loan to expand their production capacity. Her vermicomposting operation now occupies half of a 200-foot-long, 30-foot-wide (61 × 9 m) hoop house where she maintains an oval-shaped worm windrow that is 3 feet wide with 80-foot-long sides (0.9 × 24.4 m), eight Hungry Bins, a BioSafe with two drawers, and a handmade 10-by-3-foot (3.1 × 0.9 m) continuous flow-through reactor. Nancy uses a trommel with ⅛-inch (3 mm) screen and another with ¼-inch (6 mm) screen. She also has handmade tea brewers to make batches of 60 gallons and 10 gallons (225 and 38 L). The worms are fed thin sheets of fresh fruit pulp covered with finished compost.

The Worm Ladies sell worms, vermicast, worm bins, tea brewing accessories, books, and even worm earrings. They also sell aerated vermicast tea in a range of gallons—1 gallon (3.8 L) is $15 and 120 gallons (455 L) is $580.

The Worm Ladies offer workshops and clinics on vermicomposting, for which attendees can obtain Master Gardener credits or professional development credits. The state of Rhode Island has only one landfill, and it is running out of space, so Nancy emphasizes food waste diversion by vermicomposting or composting. She has written a children's book titled *What Is a Red Wiggler?* that is selling well.

Nancy also provides consulting services. She offers free assistance to home vermicomposters and charges a fee for businesses choosing to manage organic waste via vermicomposting.

Green Organic

AFGHANISTAN

Located in Kabul, Afghanistan, Green Organic Agricultural Production Company is a woman-owned organic materials recycling facility. It is the first commercial composting company in Afghanistan that has vermicomposting production onsite. Hakima Zolfaghari is the president of Green Organic, and she started the business with the goal of composting and vermicomposting 20 percent of Kabul's organic waste. Another goal was to hire and train women, so they could become skilled at vermicomposting commercially as well as at home. Hakima attended the Peace Through Business program by the Institute for Economic Empowerment of Women, which was formed to educate women who have been oppressed and their homelands ravaged by war, genocide, and poverty.

FIGURE 10.7. Green Organic is a woman-owned business in Afghanistan. *Photograph courtesy of Qasim Bakhshi.*

Hakima pitched her business plan to USAID's Assistance in Building Afghanistan by Developing Enterprise (ABADE). The ABADE folks liked that Green Organic was a sustainable, innovative, scalable project that financially empowered women, so they decided to form a public-private alliance with the business. In November 2013 ABADE outfitted Green Organic with 882 pounds (400 kg) of earthworms, small machinery, tools, and supplies. Green Organic provided farmland, funding, labor, and project management.

USAID's IDEA-NEW is a separate program that helped get Green Organic established. They provided technical assistance by helping Hakima assemble vermicomposting beds and learn earthworm husbandry.

FIGURE 10.8. Harvesting and bagging vermicast is a labor-intensive operation at Green Organic. *Photographs courtesy of Qasim Bakhshi.*

Hakima contacted me in 2015 and sent me several photos of her concrete-block vermi-beds. Azim Bakhshi works with Hakima, and we've corresponded for the past couple of years. Azim made a simple continuous flow-through vermi-bin using concrete blocks that he says is "easy to make and flexible."

They are vermicomposting fruits, vegetables, and other agricultural wastes, and they harvest vermicast after five months in the bins. Green Organic is producing 100 tons (90,720 kg) of vermicompost annually and selling it to local farmers and individuals for the equivalent of 36 US cents per kilogram. Their annual revenues are about $50,000 in US dollars.

Earthen Organics

EASLEY, SOUTH CAROLINA

Kristen and Anthony Beigay have a small but growing worm farm. They reuse and repurpose anything they can on the farm, as they have strived to build the business slowly and without taking on debt. They have plans to greatly expand their vermicomposting operation and hope to be able to supply customers in the entire US Southeast region and beyond.

They began vermicomposting as a hobby in 2012 by building a VermBin96 flow-through worm bin from plans developed by vermicomposting enthusiast Joseph Denial. Kristen was laid off from her job in the financial sector in 2013 when the market downsized and was considering her possibilities for a new job. Once the Beigays started harvesting vermicast from their worm bin, Kristen realized the potential for a vermicomposting business and decided to pursue it. Kristen now runs the company full-time with part-time help from Anthony and their two children. Anthony's mother and stepfather help out as well, so three generations are involved in the business.

The feedstock is horse manure from a local horse boarding facility. The horses are kept in stalls with pinewood shavings for bedding. The stable hands remove the mixed manure and wood shavings and put it in piles that are later loaded into Kristen's 16-foot (4.9 m) trailer.

Earthen Organics composts the manure to destroy pathogens. The manure-and-wood-chip mix composts well without any additional materials added. Kristen uses an aerated-static pile (ASP) system designed by O2Compost. She uses fresh horse manure to make a pile and covers it with old manure that has set for a couple of years. Kristen's initial compost pile contained 40 cubic yards (31 m³) of horse manure. Her current pile is 90 cubic yards (69 m³), and the next will be about 90 cubic yards. Active composting takes place for 21 days, and the composted material is cured for another 21 days. Kristen takes regular temperature readings and enters them into a spreadsheet. Peter Moon, the owner of O2Compost, provides guidance when needed.

A Worm Farmer's Inside View

I asked Kristen to tell me about some of the hurdles she and her family have faced in starting and running their business. Here's what she told me: "We started working with worms as a hobby. After I lost my job, the business possibilities came to light when we started harvesting the castings from our VB96 bin. With limited funding and the choice to make it a business, there have been numerous hurdles. Worms are finicky, and you have to give them the best conditions possible. We failed at that a few times before we finally succeeded, coming up with the right combination of bedding and feedstock. Once we got it right, we decided to try our hand at building larger bins on the premise of the VB96. The bins were great until harvest time. It is difficult to manually harvest a 4-by-8-foot (1.2 × 2.4 m) bed with garden trowels. We started going through design changes with each bed to implement a breaker bar and winch system because we had three 4-by-8 bins in operation. We didn't know that wood would expand and contract so dramatically with the moisture variations in the bin. Keep in mind we were building wood bins to keep the cost down.

"We then decided to buy design plans from someone who had years of experience and switch to an all-metal structure. We would love to say that the big bed we have now works extremely well, but we can't. We have had the breaker bar get caught in the base screen, causing it to bend both the bar and the screen. We then had to dump the entire bed to repair it and rebuild it. Now we have the breaker bar and winch system working, but we are working through inconsistencies in the bedding, which cause it to fall out of the bottom of the CFT bin, and how to repair that.

"To this day we strive to make the business sustain itself. Without loans to fund the business you learn what you need and don't need, and you base your growth on that. Not being able to afford automation of the process means a lot of hours of manual labor. Even our container loading and labeling is done by hand. Also, we are fortunate to have a building to use, but it isn't set up well for what we are doing. Those familiar with process flow charts would be disturbed by our 'spaghetti flow analysis,' but there is little we can do in our confined space. Improving that means spending a lot of money on new facilities.

"Anthony has a full-time job and then comes home to do the heavy lifting, equipment design and testing, and markets. Caleb juggles his school and work schedule so he can help at markets and around the farm. I am the only full-time worker and can't do the heavy stuff alone. We have had many trials in our growing process. We continue to improve our throughput and quality but still face struggles. Things would and could be much easier if money were no object."

After pre-composting is complete, the Beigays screen the compost before feeding it to the worms. They have a handmade 5-foot-long (1.5 m) trommel with 1-by-½-inch screen (2.5 × 1.3 cm), and an industrial drill is used as the motor. Kristen's husband, Anthony, has purchased bicycle rims to build a new trommel that will be 8 feet (2.4 m) long. Its screen will be ½ inch by ½ inch (13 × 13 mm), and it will also only be used for thermophilic compost.

The worms are housed in a 2,500-square-foot (230 m²) building that was originally constructed as a wine cellar. The 60-by-36-foot (18 × 11 m) structure is completely underground, so the temperatures are steady, dropping only as low as 52°F (11°C) in the winter, and rising only to 70 to 72°F (21–22°C) during the summer.

Earthworm breeding takes place in shallow mortar trays inside the cellar. The Beigays mix ⅛-inch (3 mm) screened horse manure compost and peat moss to make bedding for the trays. The breeding worms are fed a mixture of juice pulp and a custom worm chow recipe they create by running the ingredients through a coffee grinder. The worm offspring are intended for their vermireactor. They occasionally sell worms to a few nonprofit groups and donate worms to some others, including one that works with inner-city youth.

Thousands of *Eisenia fetida* earthworms are in a 48-foot-long by 5-foot-wide by 30-inch-deep (14.6 × 1.5 × 0.8 m) hand-built steel continuous flow-through bin. The bin has 4-by-2 inch (10 × 5 cm) grates on the bottom. To improve the bin design, Kristen says the next bin they build will have 2-by-2-inch (5 × 5 cm) stainless steel grates.

The worms are fed raw juice pulp. Kristen picks up 30 to 100 gallons (115–380 L) of pulp per week in 5-gallon (19 L) food-grade containers. The pulp consists of a variety of fruits and vegetables such as greens, carrots, apples, pears, ginger, pumpkin, beets, watermelon, celery, and sweet potatoes. They have been processed through a powerful juicer, so the pulp is dense and not very moist. The pulp contains no citrus and no fruit peels.

FIGURE 10.9. The pre-screen trommel at Earthen Organics. *Photograph courtesy of Kristen Beigay.*

The pulp is spread in a thin layer on top of the worm bed and then covered with composted manure to keep out insects and other pests. Next, they spray water to dampen the compost.

All tasks are done by hand, including feeding and watering the worms and moving the organic material. There is not enough height clearance in the cellar to have a gantry feeder above the worm bin; nor is there room for a conveyor system. They are trying to develop an overhead automated watering system but have to overcome water pressure issues. For now, they use a handheld hose to water the bin. Because the worm bin is inside a cellar, the bedding stays pretty damp and doesn't require daily watering. It takes two people an hour to feed the worms and wash out the buckets and around two hours to harvest and screen the vermicast.

The vermicomposting process takes approximately 45 days. Whenever the Beigays harvest vermicast, they collect close to a cubic yard (0.8 m³) of material. During winter and early spring when temperatures are cool, they harvest once a week or every other week. In the summer when the air temperature is 72°F (22°C) and it is 85°F (29°C) in the vermi-reactor, the worms consume a larger

FIGURE 10.10. Inside Earthen Organics' continuous flow-through bin. *Photograph courtesy of Kristen Beigay.*

FIGURE 10.11. Harvesting the continuous flow-through reactor at Earthen Organics. *Photograph courtesy of Kristen Beigay.*

amount of food. Thus, the worms are fed twice a week, and the bin is harvested twice weekly.

The vermicast is harvested by activating a motor to pull a breaker bar across the top of the wire-mesh bottom of the reactor. The vermicast falls on the floor, and they sweep it up and put it into mortar trays. The castings are screened in a designated commercially-manufactured aluminum 12-foot-long (3.7 m) long rotating trommel. First, the castings pass through a ⅛-inch (3 mm) screen, followed by a ¼-inch (6 mm) screen. The smallest particles are sold as vermicast; the larger ones are put back in the worm bed. In early 2018 they were producing 40 to 50 cubic yards (31–38 m³) of vermicast annually.

Their vermicast is approved for use at Clemson University–certified organic farms. They sell vermicast wholesale in bulk in 55-gallon (210 L) drums or by the cubic yard (0.8 m³). They package vermicast in 1-, 2-, and 5-gallon (3.8–18.9 L) plastic reusable storage buckets that may be refilled at a discount at designated retailers and farmers markets.

In 2018 they introduced a new product line—potting mixes—that is being well received by customers. One potting mix includes coconut coir, rice hulls, oyster shells, vermicast, and compost with wood chips. It is suitable for containers because of the large proportion of aeration materials—it doesn't compact. A different potting mix is heavier (less rice hulls and more compost) so it doesn't wash out easily. The organic ingredients are blended in a cement mixer (although they have plans to purchase a soil mixer). The mixes are sold in cubic-yard (0.8 m³) totes, 1½-cubic-foot (0.04 m³) bags, and ½-cubic-yard (0.4 m³) bags. They have custom-printed labels to adhere to the white plastic bags.

Kristen, along with her husband or son, Caleb, has attended several of my NC State Vermiculture Conferences. They've made the commitment because they learn new ideas and make valuable

FIGURE 10.12. Vermicast is sold in refillable buckets by Earthen Organics. The buckets are food-grade, easy to carry, reusable, and recyclable. *Photograph courtesy of Kristen Beigay.*

connections with other attendees. When Caleb first attended the worm conference, he was an aeronautical engineering major at Embry-Riddle Aeronautical University in Daytona, Florida. Halfway through the first day of my conference, Caleb got excited after hearing a speaker from Belgium who has the largest vermicomposting operation in Europe (see PUR VER later in this chapter), and he transferred to Clemson and changed his major to plant and environmental science.

Arizona Worm Farm

PHOENIX, ARIZONA

Creating a sustainable system was part of Zach Brooks's motivation for starting the Arizona Worm Farm. Zach finds local sources of wastes that he can compost and vermicompost to produce fertility for the crops he grows and to sell vermi-cast products. Right now his operation is moderate-sized; he has about 300,000 worms. His goal is to get to 2 million!

In 2013 Zachary Brooks went back to Arizona State University to get a second master's degree in sustainability. He decided to adopt an off-the-grid lifestyle using solar and wind power, and recycling organic waste. Zach bought a 10-acre (4 ha) farm on the south side of Phoenix, Arizona, at the base of South Mountain, and began leasing 20 more acres (8 ha) in December 2016.

He makes compost that he uses to feed the soil and grow fruits and vegetables. Worms, black soldier fly (BSF) larvae and frass (their manure), and food waste are fed to chickens that provide Zach's family with eggs and meat.

Zach's land is zoned for agriculture. He was not required to get a solid waste processing permit because he uses most of the compost onsite to grow food. The agricultural zoning designation includes the raising of livestock, and his earthworms fall into that category.

Zach established a business model where he gets paid to take others' waste, and he turns it into value-added products that may be used onsite or sold. A vegetable processing company delivers 40 tons (36,285 kg) of waste to Zach's site per week. He bought two 40-yard (37 m) containers and placed each one at a horse stable. When the containers are full of manure, a company delivers them to his site. He receives green waste and woody materials from landscape companies. Initially, he accepted landscape waste and hired someone to grind it. Now he only takes green waste from businesses that grind it themselves before delivery to Zach's site.

In early 2018 Zach had 10 composting windrows that were each ¼ mile long (400 m). To start a compost pile, he lays down mulch, then covers it with food waste, next horse manure, then green waste, and finally rock dust. These alternating feedstocks are laid in the windrow until it reaches the height of the John

FIGURE 10.13. Picking up horse manure to feed to worms at Arizona Worm Farm. *Photograph courtesy of Zachary Brooks.*

FIGURE 10.14. Composting windrows at Arizona Worm Farm. *Photograph courtesy of Zachary Brooks.*

FIGURE 10.15. Worm bins inside a hoop house at Arizona Worm Farm. *Photograph courtesy of Zachary Brooks.*

Deere compost turner. He lets it go through three heat cycles and then allows the compost to cure in the windrow for four to six months. Compost windrows are harvested once a year.

Zach is vermicomposting onsite in four different ways. He propagates worms in mortar trays inside a 50-by-36-foot (15 × 11 m) insulated barn with evaporative cooling. He also has two continuous flow-through reactors inside the barn. Two shade houses contain worm windrows that are 100 feet long by 4 feet wide (30.5 × 1.2 m). Zach has had good results from experimenting with the wedge system, so he is putting up more shade houses to contain 40 vermicompost wedges.

Zach pre-composts feedstock separately to feed to the worms. He creates static piles with shredded corrugated cardboard, horse manure, and vegetable waste. They go through two heat cycles before they are fed to worms.

He harvests the vermicast two or three times per year. The vermicast is dried in the sun for a day or two until it reaches 25 to 30 percent moisture level. A trommel is used to screen the vermicast. He uses the overs as mulch on the farm. As of early 2018 he was selling 50,000 pounds (22,680 kg) of castings annually.

Zach sells worms by the pound for $30; sales volume is about 25 pounds (11.3 kg) of worms per week. Vermicast is sold in 15-pound (6.8 kg) bags for $15 and

25-pound (11 kg) bags for $25. He also sells it in bulk by the cubic yard (approximately 220 gallons/835 L) or 55 gallons (210 L). Zach offers a vermicompost mix for backyard gardeners. It consists of compost, 12 percent vermicast, coir, vermiculite, and perlite. Mulch and compost are sold in 2-cubic-foot (0.06 m³) bags for $3 or in bulk for $15 per cubic yard (0.8 m³). People can come to the farm to buy the bulk products, or Zach will deliver for a $65 to $100 fee, depending on the customer's location. Arizona Worm Farm also sells small worm bins for households. The bins contain beddings, worms, and food and are sold for $40.

Zach holds 90-minute classes at the farm on one or two Saturday mornings each month. Attendees pay $60 to learn about vermicomposting and take home a worm bin. They learn how to set up and maintain a worm bin and use vermicast to enhance their organic gardens and landscapes. He also speaks to groups of 30 to 50 people weekly. His goal for 2018 is to teach 10,000 people to vermicompost at home.

Germina, Inc.

TURKEY

In 2013 five physicians and an engineer began research and development of replicable strategies for sustainable farming to promote good farming practices in Turkey. By 2017 they had established a commercial vermicomposting operation, the largest black soldier fly (BSF) larvae colony in Turkey, a flock of free-range chickens (that eat BSF larvae), and a pilot project on aquaponics. Their goal is to use these four practices as an integrated system.

They opened their production facility on a 13½-acre (5.5 ha) cattle farm in Elmadag, Ankara. It houses, on average, 350 Simmental beef and dairy cows. The cattle arrive at the farm at 9 to 13 months of age. Their high-quality feed includes hay, silage, alfalfa, maize, barley, and oats. When they are 14 months old, the cows are artificially inseminated and then sold when they are 3 months' pregnant. Their average manure output is 35 to 40 pounds (16–18 kg) per day.

The partners began construction on a production facility in February 2017 and completed it two months later at a cost of $100,000. The warehouse consists of 7,158 square feet (665 m²) of closed space separated into sections. The vermi-reactors are in one 3,650-square-foot (340 m²) area. Collection, packaging and storage takes place in a 1,550-square-foot (145 m²) room. There is a section of about 900 square feet (84 m²) for the preparation of manure feedstock, including sifting and thermal processing.

Cow manure is washed out of the barn and run through a solids separator. The manure is put in piles outdoors for one or two weeks. This allows more liquid to evaporate from the manure, and the piles heat up to around 150°F (66°C). The next step is to bake the manure at 158°F (70°C) for an hour. This

FIGURE 10.16. This motor-driven trommel with adjustable legs screens Germina's harvested vermicast in Turkey. *Photograph courtesy of Ali Calikoglu.*

step is required by the Turkish Ministry of Food, Agriculture and Livestock. The group designed a 283-cubic-foot (8 m³) oven for this purpose. The cow manure is placed in this oven in 1-inch-deep (2.5 cm) trays and processed for one hour. The next step is to sift the manure to separate it from unwanted materials such as pebbles, rock pieces, and large chunks of manure. They run it through a trommel screen with a mesh size of just under ½ inch (13 mm) that is operated by a small engine.

They designed and built 10 continuous-flow vermi-reactors for $80,000. The bins are 50 feet long by 4 feet wide by 2½ feet deep (15.2 × 1.2 × 0.8 m). The frame of the bins is cast iron; the walls are made of polyurethane material, which provides heat insulation without allowing the growth of mold and parasites. The material is also resistant to corrosion and requires minimal maintenance. There is a 1½-by-1½-inch (4 × 4 cm) iron mesh at the bottom of each vermi-reactor.

The group started with an initial stock of about 20,000 *Eisenia fetida* earthworms in December 2015. In a separate facility they propagated those worms until there were approximately half a million worms to begin vermicomposting. They estimate that in 2018 the earthworm population had increased to 3 million.

When they prepare a vermi-reactor for vermicomposting, they put four or five layers of used newspaper over the open mesh bottom and place 4 inches (10 cm) of compost on top of the paper. Worms are then added at a rate of 10,000 per 10.8 square feet (1 m²). They begin feeding the worms 0.4 to 0.8 inch (10–20 m) of manure two to three times a week. Once the worms are acclimated to the environment, the amount of feeding is increased slowly up to 1.2 to 1.6 inches (3–4 cm) per feeding.

Initially there were problems with the bins overheating. They realized that the cause was overfeeding, so they cut back on the amount of food applied to the bins.

Another challenge was not enough heat in the bins during winter. To remedy this situation, they increased the amount of food for the worms and placed heating cables on top of the worm beds.

It takes four to six months for a bin to reach a level that allows harvesting. The harvest from an individual bin is about 66 gallons (250 L).

The group installed a unique, sturdy harvesting mechanism in each bin. Just above the mesh bottom is a ½-inch (13 mm) blade attached to two thick roller chains on both ends and pulled by a 1-horsepower (0.75 kW) motor. When the harvesting begins, this blade shaves off around ½ to 1 inch (1.3–2.5 cm) of vermicast.

Conveyor belts run underneath the bed grates to catch and transport the harvested vermicast to the packaging and storing section via a central conveyor belt. The vermicast is screened to separate unintentionally harvested worms. The vermicast matures for four weeks in this unit before packaging begins.

In February 2018, 2 of the 10 vermi-reactors had reached full capacity. About 1.3 cubic yards (1 m³) of vermicast was being harvested weekly. When all 10 vermi-reactors are operating, the partners predict they will have 392 to 458 cubic yards (300–350 m³) of annual production of vermicast. They plan to sell their product in bulk to greenhouses.

Grupo Aldea Verde

Mexico

This vermicomposting operation is Mexico is so far reaching and is growing so rapidly that it is hard to describe all that they are doing. Their goal is to create environmental, social, and economic well-being through sustainable management of organic materials using the best technology available. They want to provide food security through the recovery of depleted soils.

This vermicomposting business was founded in 2005 by Aurora Niembro, who had the idea of developing a multifamily project for their retirement. She motivated her husband and two other families, the Vargas and Casillas, to

FIGURE 10.17. This palm oil extractor plant in Mexico has enormous composting windrows that are aerated with the blue-and-white compost turner. *Photograph courtesy Francisco Niembro.*

undertake the primary goal of providing services and added-value end products that promote health, environmental care, and social development.

After they received basic instructions (there were no qualified experts on vermicomposting at that time in Mexico) and bought enough worms to install their first eight worm windrows, the fun began on their 2½-acre (1 ha) piece of land. By year 4 (2009) they had 26 worm windrows, each 131 feet long by 8 feet wide (40 × 2.4 m), and they were producing over 300 tons (272,155 kg) per year of vermicast.

At that time they had to evaluate the future of Aldea Verde and finally decided to keep on growing. This is when Aurora's son, Francisco Niembro, got involved. Francisco is a dentist, and as if he didn't have enough to do tending to his patients, he decided to accept the invitation to become Aldea Verde's CEO. They bought a bigger piece of land (10½ acres/4.3 ha) in the same central state of Querétaro and set up 50 worm windrows that are each 115 feet long and 8 feet wide (35.1 × 2.4 m).

Also in 2009 they began to offer consulting for new vermicomposting projects, and Grupo Aldea Verde (GAV) started to operate as a group of businesses. After accepting Javier Vargas Jr.'s alliance and leadership, they developed Aldea Verde's second production facility, at Tamaulipas state up in the northeast region, with another 24 worm windrows there.

After installing even more private midscale vermicomposting projects, GAV developed a vermicomposting model called Social Projects. By 2018 they created 13 vermicomposting facilities that provided improved employment and income to over 100 families in seven marginalized communities in three Mexican states. GAV recruited 50 partners (from academic, research, private, and public sectors, as well as community service organizations) to help with this outreach. The Social Projects are transforming more than 3,000 tons (2.7 million kg) of organic waste into 1,000 tons (907,185 kg) of vermicast, 132,086 gallons (500,000 L) of tea, and 1.1 tons (1,000 kg) of worm flour per year.

In 2013 a joint venture was achieved between Grupo Aldea Verde and Compostamex, whose chief engineer is David Escalera, a highly experienced composting expert in Mexico. Escalera and Francisco Niembro decided to work together on the development and promotion of new mid- and large-scale composting and vermicomposting facilities. Mexico's national environmental agency asked them to help create national composting and vermicomposting standards.

In 2015 Compostamex and Aldea Verde established another joint venture with composting expert Peter Moon from O2Compost, which is located in Snohomish, Washington, in the United States. Together they are promoting aerated composting systems in Mexico.

By 2018 Francisco and his partners had developed more than 70 vermicomposting operations in Mexico and other Latin American countries. The projects they have helped establish are processing more than 5,000 tons (4.5 million kg) of organic materials per day. Here are brief descriptions of some of the projects they have installed:

A tequila producer is composting 300 tons (272,155 kg) per day of agave bagasse and will begin vermicomposting in 2019.

The most productive palm oil extractor plant in Mexico has been composting 15 tons (13,610 kg) per day since 2014, and by April 2018 they expanded to 80-plus tons (over 72,575 kg) per day. In 2019 they will begin vermicomposting with 20 continuous-flow reactors, each 262 feet (80 m) long, and will produce an estimated 4,000 tons (3.6 million kg) of vermicast per year.

The second largest meat producer in Mexico has a 140,000-head cattle lot that generates 500 tons (453,590 kg) per day of manure, hay, and mortality. They have been composting since 2015, and in 2018 they are expanding by reengineering their composting facility and adding vermicomposting.

A major berry grower and packer is vermicomposting 10 tons (9,070 kg) a day of their packer residues and producing 27,740 gallons (105,000 L) of vermicast tea per week. They vermicompost in 55-gallon (210 L) drums cut in half lengthwise and laid on their sides.

GN Productores Agricolas grows 7,413 acres (3,000 ha) of cabbage, brussels sprouts, onions, berries, and other crops. They process 50 tons (45,360 kg) of waste per day from two food packers and produce compost, vermicast, and teas for their own use, which has increased production over 30 percent.

Moon Palace Hotel in Cancun is processing 5 tons (4,535 kg) per day of their own food scraps and yard waste using in-vessel composting and vermi-composting. They close the loop by using the end products of compost and vermicast to grow vegetables for the hotel's restaurants.

A large-scale fruit producer in Tecoman, Colima, processes 20 tons (18,145 kg) per day of papaya and pineapple waste to produce compost, vermicast, and teas for its own use, for the goal of increasing crop production and using less synthetic fertilizer.

In early 2018 Aldea Verde was processing 5,000 tons (4.5 million kg) per year of cow, horse, and sheep manures in 74 outdoor windrows in Querétaro and Tamaulipas, Mexico. They propagate worms to start new projects and sell more than 1,800 tons (1.6 million kg) of vermicast yearly and 34,315 gallons (129,900 L) of vermicast tea monthly. They are building 60 continuous flow-through reactors 131 feet long (40 m) to increase production to over 8,000 tons (7.3 million kg) a year of vermicast in 2019.

FIGURE 10.18. PUR VER's Alexandre Meire in Belgium. *Photograph courtesy of Alexandre Meire.*

PUR VER

BELGIUM

Pur ver means "pure worm" in French. Alexandre Meire is the founder and managing director of PUR VER SA in Belgium. Dr. Grégory Sempo is in charge of research and development. Their offices are in Gembloux, and their vermi-production facility is located in Pecq (near Tournai), on a farm belonging to Stéphane Cossement.

Meire began working with faculty at the Agro-Bio University in Gembloux in 2012 to develop a unique vermicomposting system. With public and private funding, they designed and built two prototypes of

FIGURE 10.19. PUR VER has highly engineered continuous flow-through bins in Belgium. *Photograph courtesy of Alexandre Meire.*

continuous flow-through reactors. They developed a suitable system and then had to find a place to set it up for commercial production. A chance meeting led Meire to Stéphane Cossement, who is renting a shed on his farm for the vermi-operation and providing labor and equipment. The first two production lines of continuous-flow reactors were constructed in 2013 and became operational in 2014. The reactors are 131 feet long and 8 feet wide (40 × 2.4 m). In 2015 four more production lines were added after five members of the Business Angels Netwerk Vlaanderen contributed €240,000.

They maintain a high density of earthworms in six lines of automated production. PUR VER produced over 250 tons (226,795 kg) of vermicast in 2017. Their ultimate goal is to generate more than 1,000 tons (907,185 kg) per year.

They vermicompost vegetable matter, coffee grounds, and spent grain in a controlled process to ensure high-quality and consistent products. The feedstock is agro-industrial waste that has already been composted or processed. Their vermicast's use is authorized in organic farming according to the European Commission Regulation 834/2007.

Meire jokes that his enterprise is the largest employer in the region, with millions of workers (*Eisenia fetida* worms). He says, "That is true that we do not pay them, but we must house them, provide them heat and food, and ensure their good working conditions!"

The team takes a scientific approach to produce the vermicast and market it. They collaborate with several scientific partners such as universities and research centers to test their vermicast's effect on soil and plants. They use the results to market to specific customers and recommend certain quantities to use.

The company sells vermicast to the general public at garden centers in 44-pound (20 kg) bags, 6.6-pound (3 kg) buckets, and 2.75-pound (1.3 kg) bags. Vermicast is sold in 1,102-pound (500 kg) bags to professional customers such as market gardeners, municipalities, park contracts, and gardens.

WormCycle

PLYMOUTH, MICHIGAN

Gavin Newton is a commercial airline pilot who used to fly with Jack Chambers, the owner of TerraVesco, which is described later in this chapter. During their long flights, Jack described how much he enjoyed worm farming, and Gavin decided that he had to try it. He moved to the Detroit area over six years ago and bought a 20-acre (8.1 ha) homestead. Gavin converted an existing garage and now has a 150-foot-long (46 m) building to house his vermicomposting operation. He decided to call his business WormCycle to symbolize feeding the worms, the worms helping the soil, and the soil helping to feed us: "The Worm Cycle is complete," says Gavin.

Gavin believes in repurposing used materials, so he experimented with and built his own equipment from materials he had on hand. This included different versions of manual screeners and a continuous flow-through bin (CFT) made out of wood. The reactor worked well at first, but eventually the forces exerted during the harvesting process overwhelmed it. The reactor collapsed, leaving Gavin with a pile of vermicast and worms on the floor. Gavin had to figure out another bin setup for the worms, because it is too cold during a Michigan winter to leave them in piles on a concrete floor. Fortuitously, a customer named Dan Lonowski came by to buy some vermicast, and Dan told Gavin he was about to start a company called Michigan SoilWorks and begin manufacturing continuous flow-through bins. Gavin bought two of Dan's CFTs. Both are 4 feet wide by 30 inches deep (1.2 × 0.8 m); one is 16 feet (4.9 m) long, the other 8 feet (2.4 m).

Gavin's building is not heated, so he built a "room within a room" around the worm bins. The wall frames are stuffed with insulation that keeps the space around the worm bins over 50°F (10°C) during winter.

Gavin drives his pickup truck a mile down the road to racing stables to pick up horse manure. The manure is in piles that have been through thermophilic heating cycles; therefore it is stabilized and sufficiently aged and may be fed directly to worms. The horse manure is a couple of years old before it is fed to the worms.

From the pickup bed, Gavin shovels the manure into wheelbarrows. Then he wheels them to the worm bins and shovels the manure onto the top of the

FIGURE 10.20. A continuous flow-through bin manufactured by Michigan SoilWorks. *Photograph courtesy of Dan Lonowski.*

beds. We joked about how he doesn't need a gym membership with shoveling so much heavy manure. He's brainstorming ways to mechanize a system to load a hopper on the ground and lift it up onto the top of the worm bin.

Gavin captures rainwater to moisten his worm beds. During winter, he uses snowmelt for wetting the bed with a 2½-gallon (9.5 L) watering can. This is in keeping with the ethos of using the materials he has on hand.

The *Eisenia fetida* earthworms in Gavin's bins process the horse manure in 60 days. During winter, Gavin harvests the bins once a week, and during summer, twice a week. The 8-foot-bin has a hand crank (like you would use to pull a boat up on a trailer) to pull the breaker bar. The 16-foot bin has a bidirectional motor that moves a single cutting blade, affixed to a flanged carriage that rides on a guide rail, like a train on its tracks. The blade travels about an inch (2.5 cm) above the grating and is moved back and forth by a continuous loop cable. The return loop is below the bin.

The harvested vermicast falls onto the floor beneath the bin. Gavin lets the vermicast dry out to 30 percent moisture, because if it is too moist, it will clump up and not go through a screen. Gavin uses a broom to sweep the castings out from under the bin, and then he shovels the castings into a wheelbarrow. The screen system that Gavin built himself proved to be too slow at processing casting, so he bought a Brockwood Worm Shi*fter, which has three screen sizes. Gavin sells the smallest particles and puts the larger ones back in the worm beds. In early 2018 he was producing 3 cubic yards (2.3 m^3) of vermicast per month year-round.

Sales of vermicast slow down during winter, then pick up in March. Spring and fall are his busiest sales seasons. He sells vermicast to a couple of local greenhouses, and orders through his business's website are increasing. He puts labels on cubic-foot sandbags to sell his vermicast. The cannabis cultivating industry is growing substantially in Michigan, so Gavin foresees that market opening up for his vermicast.

Gavin has plans to expand his operation. His current building has room to house four 40-foot (12 m) continuous flow-through bins, and he would like to build up to that and eventually make further expansions. Jack Chambers gave Gavin sound advice, saying there are no shortcuts. You can't just throw a bunch of money into a vermicomposting operation; you have to grow it over time.

Mass Wiggle

Petaluma, California

What happens when a master mechanic, an internet advertising salesman, and a third-generation dairy farmer decide to start a worm farm? You get a creative approach to maximizing space efficiency. Mass Wiggle is co-owned by Glen Ghilotti, David Bailey, and Don Silacci and located on Don's 150-acre (61 ha) farm near Petaluma, California. In 2013 they set up a vermicomposting operation, making most of their equipment themselves.

Hand-built continuous-flow reactors are stacked *three-high* inside a completely enclosed 160-by-60-foot (49 × 18 m) barn. This is a unique vermireactor design I'd not seen before. There are 24 bins that are 70 feet long by 8 feet wide by approximately 2 feet deep (21.3 × 2.4 × 0.6 m). The bins are arranged in four rows with six bins per row. The Mass Wiggle owners installed over 1 mile (1.6 km) of ½-inch (13 mm) water hoses for misting the bins.

Every bin is covered with erosion control cloth, which is very durable. They bought 150-foot (46 m) rolls of erosion cloth for $2,500 apiece. The purpose of the erosion control fabric is to prevent moisture from evaporating from the bins. Plus, the fabric is sturdy enough to support wet, heavy vermicast that has been harvested from the bin above and dropped down onto the bin below. They rake vermicast off the tops of the erosion cloth.

FIGURE 10.21. Mass Wiggle's continuous flow-through bins are stacked three-high to maximize the vertical and horizontal space inside a barn in Petaluma, California. *Photograph courtesy of Will Bakx.*

FIGURE 10.22. This sand bucket side-shooter can be attached to a Caterpillar skid steer. The operator scoops up compost in the bucket; as the augers turn, they propel the compost onto the conveyor belt below the bucket. The belt carries the compost to either side for dispensing into worm bins. *Photographs courtesy of Will Bakx.*

FIGURE 10.23. This conveyor moves harvested vermicast into a hand-built trommel screener at Mass Wiggle. *Photograph courtesy of Will Bakx.*

The farm has 50 dairy cows. Their manure is collected with a loader and mixed with straw, wood shavings, and a small amount of horse manure (5 percent of the total). The mixture is composted in windrows for about 45 days before it is vermicomposted.

A sand bucket side-shooter attached to a Caterpillar skid steer is used to pick up the compost. Two augers inside the bucket drop the compost onto a conveyor belt built into the bottom of the bucket. The conveyor can discharge the compost from either side of the bucket to feed the worm bins. The skid steer drives between the rows and spits feedstock into the beds, and then workers rake the feedstock out evenly by hand. The worms are left alone for the rest of the week after they are fed.

Mass Wiggle uses multiple loaders and tractors for each separate task to prevent cross-contamination of pathogens.

It takes approximately 180 days for the vermicomposted material to go from the top of the bin to the bottom to be harvested. They cut 1 inch (2.5 cm) of vermicast off the bottom every week, so it is 20 weeks (almost half a year) for the approximately 24-inch-deep (61 cm) bin to empty. They harvest 24 cubic yards (18 m³) per week.

They purchased four different types of conveyor belts for post-production handling of the vermicast. Harvested vermicast is spread on a concrete floor next to the screening system to allow excess moisture to evaporate. When the vermicast has reached a sufficient moisture level, it is loaded into a hopper. A conveyor delivers the castings to a trommel that was built by Glen and Don. The interior screen is ¼ inch (6 mm); the exterior screen is ⅛ inch (3 mm). The screened vermicast drops down onto a plywood ramp and goes to a conveyor. Then it travels up another conveyor to the bagging system. All of this takes place inside a huge barn. Overs are removed and spread on a back field where silage is grown.

Mass Wiggle doesn't sell earthworms, because all of their worms are needed for vermicast production. When I visited Mass Wiggle in 2017, they had recently begun bagging their vermicast. They decided to use paper bags with a plastic inner liner. They do batches of bagging and date the bags. Each bag holds ½ cubic foot (0.01 m³) of vermicast and sells for $20. Mass Wiggle is certified organic by the California Department of Food and Agriculture.

TerraVesco

Sonoma, California

TerraVesco (Latin for "feed the soil") got going in 2012, but it grew out of Sonoma Worm Farm (SWF), which was established by Earl Schmidt in 1970 to sell worms and vermicompost to local farmers and gardeners. In 1992 a commercial airline pilot named Jack Chambers bought 5 pounds of worms from Earl, and he was so impressed by how quickly they turned his compost into rich vermicast, Jack convinced the owner to sell his house, farm, and vermi-business to him. Jack experimented with vermicomposting in his spare time and slowly grew the business, but once he retired in 2011 after 33 years as a pilot, he was able to devote more time to his worm farm. Jack adapted a three-bin forced-air static pile composting system to destroy pathogens and weed seeds before feeding dairy manure to his worms. It took a lot of experimentation, but he finally developed the optimal schedule for using motorized blowers to inject oxygen and release gases, and for turning the piles and removing the compost from the bins. Jack applied his pilot's experience of utilizing rigorous operating procedures to run his worm farm and make technical improvements. He also created, and patented, a unique mechanized continuous flow-through bin to harvest lower layers of vermicast. Over several months the pre-composted manure goes from the top of the bins to the bottom. His worms are fed twice a week on the top of the bins, and vermicast is harvested once a week from the bottom using automated breaker bars. He focused on selling his vermicast to vineyards at $1 per pound. Vineyards can lose up to 20 percent of their new plantings, but Jack's customers report that they lose less than 1 percent. Jack applies vermicast and vermicast tea to his own vineyard, and out of the 400 vines he planted, only 2 didn't survive.

By 2012 Jack was producing over 200,000 pounds (90,720 kg) of vermicast annually. For a brief period he manufactured and sold the composting and vermicomposting systems he designed. But it was costing him too much out of pocket to grow the business, and he was contemplating next moves. A little later that year, Brian O'Toole approached Jack about investing in Sonoma Worm Farm after seeing PBS television's *Growing a Greener World* host Joe Lamp'l interviewing Jack. Brian and his friend Richard North invested $1.2 million to finance new continuous-flow reactors and equipment to create TerraVesco with 49 percent ownership of the company. Richard is the CEO and manages business operations, Brian O'Toole is the chief strategist for expansion, and Jack Chambers is in charge of production. They created an Academic/Scientific Advisory Board and a Business Advisory Board (which includes a former secretary of the US Department of Agriculture).

When I visited in 2017, TerraVesco had 16 employees and about 38 million worms. The worms were fed a mixture of dairy manure and rice hulls that were pre-composted to remove pathogens and weed seeds and turn it into a homogenous material. TerraVesco maintains stringent quality assurance procedures to guarantee consistent products.

Because of Jack's partnership with Brian and Richard, his operation grew 15 times larger in three years. In 2017 they had 12 aerated static bins in which they pre-composted the dairy manure for several weeks. Jack has 12 worm bins that are 60, 70, or 80 feet (18–24 m) long and 23 bins that are 130 to 140 feet (40–43 m) long. All of the continuous flow-through bins are 5 feet wide and 2 feet deep (1.5 × 0.6 m). He thinks 8-foot-wide (2.4 m) beds are too wide and may heat up because they contain so much material. Jack created innovative new ways to feed the worms and harvest vermicast. They process 100 cubic yards (77 m³) of dairy manure per week and produce 3,000 cubic yards (2,295 m³) of vermicast annually. With projected expansion, the TerraVesco team expects to generate 11,000 cubic yards (8,410 m³) of vermicast in 2019.

TerraVesco sells vermicast in 1-cubic-yard (0.8 m³) and 1-cubic-foot (0.03 m³) bags, and vermi-extract in 250-gallon (945 L) totes. Their customers include vineyard operators, organic farmers, horticulture professionals, berry growers, nut tree farmers, landscapers, cannabis growers, and home gardeners who want to grow high-quality plants using less chemical fertilizer. TerraVesco's customers have reported that they can maintain crop yields using 25 to 50 percent less chemical fertilizer. Their products are being studied in academic scientific trials comparing their crop yields to those with conventional fertilizer applications.

Noke Ltd.

New Zealand

Noke Ltd. has a massive vermicomposting business operating at sites throughout New Zealand and other parts of the world. The company is diverting industrial waste from landfills and restoring land left barren by harvesting forests.

Noke means "earthworm" in the native Maori language of New Zealand. The founder and a director on the board of Noke Ltd. is Dr. Michael Quintern. He grew up on a farm, studied agriculture and soil science in college, and has over 20 years of experience in land utilization of organic wastes on farmland. This led to his interest in creating large-scale vermicomposting operations to process industrial waste. Michael designs vermi-systems for different waste streams, and provides consulting expertise for using vermicast to enhance soils. Max Morley is the CEO and executive chair of the board of directors.

The dairy and timber industries are the principal commerce in New Zealand's central North Island region. The organic wastes from milk processing and pulp and paper factories have traditionally been disposed of in landfills. However, Noke Ltd. is vermicomposting to turn much of the waste into valuable products. Tens of thousands of tons of dewatered waste activated sludge and dissolved air flotation sludge from milk processing plants are blended with pulp and paper mill sludge to reach the proper carbon-to-nitrogen ratio and pH for vermicomposting.

FIGURE 10.24. Inputs including milk processing wastes and paper wastes are mixed before being transported to windrows. *Photograph courtesy of Max Morley.*

Noke Ltd. has billions of earthworms processing more than 220,000 tons (200 million kg) per year of organic waste at worm farming operations in Taupo, Tokoroa, Maketu, and Pinedale (Putaruru) in New Zealand. Their largest worm farm at the Pinedale site in Putaruru covers 62 acres (25 ha). Noke has for many years operated certified organic worm farms that cover an additional 62 acres, plus two smaller sites in Tolaroa. At these farms, they compost and vermicompost pulp mill solids and kiwi fruit waste. On 1¼ acres (0.5 ha) in Maketu, they mix sediments from an oxidation pond with pulp mill solids to feed worms. In Taupo, Noke operates a 22-acre (10 ha) revolving site where they blend a variety of community-based resources with externally sourced carbon bearing material.

Quintern designed an efficient vermicomposting windrow system that minimizes labor and processing time. They have been operating the windrow systems on an industrial scale for more than a decade. Their vermicomposting windrows are up to 787 feet long and 79 feet wide (240 × 24 m). The fibrous structure of the pulp mill solids has a high water-holding capacity, which allows for the operation of the worm farms without irrigation or coverage of the windrows.

The vermicomposting windrows are rotated to different areas of the farm to prevent an accumulation of nutrients in the underlying soil. When vermicast is harvested, a crop with high nutrient demands (such as maize and trees to establish new plantations) is planted where the windrow used to be to utilize the residual vermicast.

FIGURE 10.25. Viewed from above, the scale of Noke's vermicomposting farms is breath-taking. *Photograph courtesy of Max Morley.*

Noke Ltd. sells *Eisenia fetida* earthworms for vermicomposting. They are careful not to stress the worms, avoiding common harvesting techniques such as rolling them in trommel screens or exposing the worms to bright light to force them through horizontal screens. Instead they harvest and ship the worms in their bedding, which also decreases the shock of being in a new worm bed. Worms are counted in a defined area of 8 by 8 inches (20 × 20 cm) in a windrow and calculated using GIS mapping and geo-referenced aerial images.

They also sell a mixture of earthworms for soil inoculation in pastures or new orchards. The worm species are *Aporrectodea trapezoides* and *Lumbricus terrestris*, mixed with a smaller amount of *Eisenia fetida*.

Additional products for sale are MyNOKE organic certified vermicast, premium organic certified vermicast, vermicompost, organic certified aged bark mulch, and the Hungry Bin for home vermicomposting.

California Soils

VERNALIS, CALIFORNIA

California Soils is an immense facility that can supply any size order of worms, vermicast, and other soil amendments. Business is booming, and it appears to be a clean and organized operation that employs 30 workers.

Conor Davis is the sole owner of this 162-acre (66 ha) organics recycling and soil amendments production facility in Vernalis, California. His father,

Jim Davis, and a partner, Mario Travalini, started vermicomposting at this site on an old military airfield in 1993, and called the business American Resource Recovery. In the late 1990s Mario changed the name to Pacific Landscape Supply. They had worm windrows that were 3 feet high by 3 feet wide (0.9 × 0.9 m) and up to 1,200 feet (365 m) long. They fed the worms paper pulp from a nearby corrugated cardboard recycling operation. (As paper is recycled, the fibers get shorter each time until they are too small to use again to make paper.) The waste material from the recycling plant consists of small particles that contain nitrogen from the glues used to sandwich together corrugated cardboard. The paper sludge proved to be an excellent feedstock for Davis's location. This huge site is exposed to constant winds, harsh sun, and freezing temperatures during winter. The sludge forms a crust on the surface of the windrows that insulates the piles from temperature swings and protects the worms from the light. The crust holds in moisture, yet still allows for oxygen and carbon dioxide exchange, which is critical for the worms' survival.

Conor took over his dad's business in 2003 and changed the name to California Soils. Conor had 50 acres (20 ha) of vermicomposting windrows when I visited his operation in 2011. I returned in August 2017 to tour the site, and Conor had about 30 acres (12 ha) of active vermicast production using a different method that takes less space. They are feeding ground green waste and corrugated cardboard recycling pulp (paper sludge) to the worms. Conor said the paper fiber feedstock is better than green waste. The vermicomposting process is faster with the paper waste because the worms like it so much. And the vermicast test results are superior for the paper sludge feedstock than the horse bedding and green waste feedstock.

Conor also gets horse bedding from University of California–Davis and adds a small amount (5 percent of the total) to the worm feed. The horse bedding consists of manure, straw, and wood shavings, but Conor says his worms won't eat straw and shavings until they are composted. He composts immense amounts of green waste and horse bedding separately on his site.

Conor created a new method of vermicomposting in piles he calls bunkers. A new bunker is 3 feet (91 cm) tall; when it is finished vermicomposting, it has subsided to 1½ to 2 feet (46–61 cm). The bunkers are 40 feet (12 m) wide and 800 feet (245 m) long. A loader bucket of worms is added to a bunker when it is started. Feedstocks (sawdust, ground-up wood chips, and paper pulp) for the worms are pre-composted. Conor says the C:N ratio is 200:1, so it doesn't heat up because there is so little nitrogen. It takes 9 to 12 months for the bunkers to produce vermicast. During three months in the summer and three months in the winter, it is too hot or too cold for vermicomposting, so they just try to keep the worms alive; the worms are not actively vermicomposting. Conor sprays vermicast tea on the bunkers to help break down the feedstock and attract worms. He alternates spraying tea two times a week for one week and once a week the next. Each bunker yields 700 cubic yards (535 m³) of vermicast, which is 35 percent of the feedstocks fed to the worms. They have 11 or 12 rows, so they harvest a total of 8,000 cubic yards (6,115 m³) of vermicast.

The California Soils crew bags vermicast, compost, and potting mix for sale. Their garden mix contains composted forest humus, sphagnum peat moss, perlite, pumice, lava rock, compost, vermicast, humic shale ore, dolomite lime, azomite, soy meal, and gypsum. They sell two grades of vermicast, premium and standard. The standard grade contains white chunks, which are pieces of packing tape and labels. The premium vermicast is screened to ⅛-inch (3 mm) particle size. Conor proudly showed me his new, gigantic trommel screener that was 63 feet long by 8 feet wide (19.2 × 2.4 m); the top was 20 feet (6.1 m) off the ground.

The Worm Farm

Durham, California

A roadside sign that read, MAKE MONEY SELLING WORMS inspired Mark and Arlita Purser of Durham, California, to try it out. They were trying to come up with a less labor-intensive endeavor for Arlita's elderly father and thought that raising some worms might be a good idea. Little did they know that their spur-of-the-moment decision would eventually lead to the establishment of The Worm Farm in 1994.

Without knowing anything about how to care for worms, Mark and Arlita bought 100 pounds (45 kg) of worms and stuck them in their bathtub. That night a storm hit, causing the worms to flee, and they all died. Arlita still recalls how she got up during the night and, in the dark, barefoot, she stepped onto the bathroom floor and discovered it was covered in thousands of squirmy worms. But they were not deterred and purchased more worms—after they educated themselves about how to care for them. The business started to grow and then increased exponentially after Proposition 215 passed in November 1996 in California to allow the use of medical marijuana. In 2012 Mark and Arlita partnered with their son John, and their operation has grown considerably since then.

Although The Worm Farm is located in a region with a relatively mild climate, temperatures can reach over 100°F (38°C) during summer; in the winter they get cold breezes and occasional snow. Despite these temperature swings, they have successfully raised earthworms in outdoor windrows since 1994, beginning with 3,500 feet (1.1 km) of windrows (about ⅔ mile).

The Worm Farm starts a new windrow on bare ground with fresh dairy manure 6 feet (1.8 m) wide and 6 inches (15 cm) deep. They let it sit for two

217

weeks before adding earthworms, because the manure will go through a heat cycle and then cool off. They space their windrows 20 feet (6.1 m) apart, cutting a V into the soil between them to collect runoff. Drip irrigation lines are laid on top of each windrow to add moisture.

To feed the worms, they use a PTO-driven feed tractor wagon that holds 6 cubic yards (4.6 m³) of dairy manure. They drive it between the rows, and it side-loads the windrows with cow manure. This system can feed all the windrows in one day using two workers. Each windrow receives 450 pounds (205 kg) of manure every two weeks.

Predators that eat their worms include tens of thousands of robins and other birds (egrets and blackbirds), moles, opossums, and skunks. They have tried a variety of techniques to ward off predators, and some work better than others. Live traps are used to catch and relocate mammals. Mice and rats love worms, but they are not a problem at The Worm Farm because of resident cats and owls.

By 2017 the farm had 5 acres (2 ha) of land covered with 12 windrows that were 300 feet (91 m) long and 6 to 8 feet (1.8–2.4 m) wide, plus four windrows that were 125 feet (38 m) long. They harvested windrows once a year and removed 2,000 to 3,000 cubic yards (1,530–2,295 m³) of vermicast from the windrows.

For harvesting in the early days, The Worm Farm used a 10-foot-long (3 m) trommel screen on a 10-foot-wide metal frame that straddled the windrow. Workers shoveled vermicast by hand into the trommel. They handmade the trommel using 19-gauge wire with a ⅜-inch (9.5 mm) screen and motorcycle tires. The trommel had extra-large tires to help it move through mud.

When they harvest vermicast, it is thick and mudlike, so they let it dry out for six to nine months in a pile that they turn once a month using a front-end loader with a 6-cubic-yard (4.6 m³) bucket. When the vermicast has dried sufficiently, they screen it in a double trommel inside a large barn that used to house chickens. Then the vermicast is bagged or mixed with other organic materials.

Although it is a year-round operation, The Worm Farm's heavy sales season begins in January and stays strong until July. April is extremely busy, and May is their peak sales month. They have eight employees, five in the field and three in the office. They sell 100 to 200 pounds (45–91 kg) of worms (sometimes 500 pounds/225 kg) per week. They sell small quantities to individuals and drop-ship worms for 15 catalog companies. They also sell up to a dozen pounds per week via direct orders through the internet. They ship worms worldwide and sell soil mixes in six western states. Vermicast is sold in bags or in bulk.

The Worm Farm harvests earthworms two or three times a year. They also buy all of the worms and castings from nearby worm farms, in order to keep up with demand. There are nine commercial worm farms within 20 miles (32 km) that have at least ten 100-foot-long (31 m) outdoor windrows.

The Worm Farm sells three species of earthworms: *Eisenia fetida*, European nightcrawlers shipped from Florida, and *Dendrobaenas* from Georgia. They ship worms in breathable bags that are clipped shut with a zip-tie tool. They pay a worker $1 per bag to cut Agribon cloth and sew bags in 1-, 2-, and 5-pound (0.5–2.3 kg) sizes.

However, "worm and vermicast sales are the tip of the iceberg," says Mark Purser. They sell over 100 types of soil amendments and allow customers to customize their mixes. Worm Farm customers remark that they like being able to make purchases from one place. Many buyers arrive with soil amendment recipes in hand, ordering specific quantities of compost, vermicompost, coco coir, rice hulls, vermiculite, peat moss, bat guano, volcanic rock, dolomite, volcanic rock, azomite rock powder, glacial rock dust, perlite, greensand, and cow or chicken manure. The company also sells a variety of meals, including fish, bone, blood, crab, feather, and kelp. They mix 400 to 500 cubic yards (305–380 m³) of organic materials daily during their busy season. They mix everything with front-end loaders on concrete pads. They mix the day before they ship because they have learned from experience that customers may change their mind after placing an order, so they wait six days before mixing the materials.

FIGURE 10.26. A pile of 1-cubic-foot (0.03 m³) bags of castings on a stretch plastic wrapper turntable at The Worm Farm.

In 2008 the Pursers founded a nonprofit organization called The Worm Farm Learning Foundation. They teach people of all ages the health and ecological benefits of enhancing soils with vermicasts and earthworms. Many schools, scout clubs, and local college students come to the farm for field trips to learn in a classroom setting and then explore worm windrows and compost piles outdoors. Mark also teaches a one-day commercial worm farming class; the fee for the session is $1,000.

FIGURE 10.27. A front-end loader at The Worm Farm is mixing vermicast with other organic materials for shipment to a customer.

The Worm Farm business is so successful that they generate several million dollars annually in sales. And that brings us full circle: The Worm Farm is the vermicomposting operation that was featured on CNBC's *Blue-Collar Millionaires* television show in 2015.

Resources

Author's Composting and Vermicomposting Publications

For a full list of my factsheets and publications, visit the NC State Extension composting website at composting.ces.ncsu.edu. Here are some key resources I have written that cover the basics of both composting and vermicomposting:

Worms Can Recycle Your Garbage

content.ces.ncsu.edu/worms-can-recycle-your-garbage
> An extension factsheet covering the basics of how to set up and maintain a small worm bin.

Vermicomposting: A 5th Grade School Enrichment Curriculum

content.ces.ncsu.edu/vermicomposting-a-5th-grade-school-enrichment
-curriculum
> A vermicomposting curriculum that can be adapted for any age group.

Large-Scale Organic Materials Composting Guide

content.ces.ncsu.edu/large-scale-organic-materials-composting
> A guide to the thermophilic composting of large-scale organic materials.

Backyard Composting of Yard, Garden, and Food Discards

content.ces.ncsu.edu/backyard-composting-of-yard-garden-and-food-discards
> A guide for beginners to small-scale backyard thermophilic composting.

Online Compost Recipe Calculators

These calculators will help you develop a recipe for pre-composting feedstocks for your worms:

Cornell Composting: compost.css.cornell.edu/download.html
Green Mountain Technologies: compostingtechnology.com/resources
/compost-calculator

Michigan State University: www.canr.msu.edu/managing_animal
  _mortalities/composting_tools
US Composting Council: compostingcouncil.org/wp-content/uploads
  /2014/02/Compost-Calculators-v.2.3.xls
Washington State University: puyallup.wsu.edu/soils/compost
  -mix-calculator

Persistent Herbicides

Certain herbicides can remain active in hay, straw, manure, and grass clip-pings—and they can even survive composting and vermicomposting. Read the information on these links so you understand the problem and how to avoid it:

Herbicide Carryover in Hay, Manure, Compost, and Grass Clippings (NC State Extension)

content.ces.ncsu.edu/herbicide-carryover

Persistent Herbicide FAQ (US Composting Council)

compostingcouncil.org/persistent-herbicide-faq

Clopyralid in Compost (Washington State University)

puyallup.wsu.edu/soils/clopyralid

Fuel Delivery and Containment Plan

In case you will have fuel delivered or stored on your site, this document provides guidelines for safe management practices:

Aboveground Petroleum Tanks (Purdue University Extension)

www.extension.purdue.edu/extmedia/PPP/PPP-73.pdf

Laboratory Testing of Vermicast, Compost, and Feedstocks

Laboratory testing can provide important insight into the composition and safety of vermicast, compost, and feedstocks. These resources will help you navigate the testing process:

Waste/Compost Analysis (North Carolina Department of Agriculture and Consumer Services)

www.ncagr.gov/agronomi/uyrwaste.htm
  A guide to preparing samples for submission and interpreting waste/
  compost analysis reports.

Compost Lab List (North Carolina Department of Environmental Quality)

files.nc.gov/ncdeq/Compost%20Labs.pdf

A list of private laboratories in the US that analyze compost and vermicast samples.

Labs (US Composting Council)

compostingcouncil.org/labs

A list of laboratories that participate in the US Composting Council Seal of Testing Assurance program.

Business and Marketing

US Small Business Administration (SBA)

www.sba.gov

The SBA has a sizable amount of information on their website about starting a business and writing a business plan. After reading all of the material, check out the SBA's online video tutorials on a variety of topics. In addition, the SBA has local offices throughout the United States. This means you can meet with an SBA counselor for advice and guidance, and they even offer free classes and webinars. It is truly amazing the abundance of assistance you can receive at no cost.

Choosing a Business Name

www.sba.gov/business-guide/launch-your-business/choose-your-business-name

The US Small Business Administration website explains how to choose a business name and protect it by registering with state and local agencies.

Sources of Loan Funding

The USDA Farm Service Agency provides low-interest farm loans. See if you are eligible for their Micro Loan Program, Direct Farm Operating Loans, or Intermediary Relending Program at www.fsa.usda.gov/programs-and-services/farm-loan-programs. Here are some other sources of loans provided by the USDA:

- **General Grants and Loans:** www.nal.usda.gov/afsic/grants-and-loans-farmers and www.usda.gov/topics/farming/grants-and-loans
- **Minority and Women Farmers:** www.fsa.usda.gov/programs-and-services/farm-loan-programs/minority-and-women-farmers-and-ranchers/index

- **Veteran Farmers:** newfarmers.usda.gov/veterans

BeginningFarmers.org also has information about grant and loan funding sources at www.beginningfarmers.org/funding-resources.

Websites and Online Promotion Services

Weebly (www.weebly.com) and Yola (www.yola.com) are two website-building platforms that you can use for free. They provide templates, tutorials, and other tools that make it easy for you to set up a website.

Google has developed a variety of free online products that can help you with marketing and promotion. Here are some examples:

- **Google Analytics:** An analytics platform that provides information about visitors to your website.
- **Google Trends:** A tool that can tell you how many people are searching for worms and vermicast in your area.
- **Google AdWords:** An online ad campaign platform. Some worm farmers have said that AdWords is the best tool they've used for advertising. One farmer said that he gets 99 percent of his traffic from this tool.
- **Google My Business:** A platform that provides a free listing for your business to develop visibility in Google Search and Google Maps.
- **Google AdSense:** A tool to raise money by placing ads on your website from companies with related products.

Miscellaneous Resources

NC State Annual Vermiculture Conference

composting.ces.ncsu.edu/vermiculture-conference

Learn directly from worm farmers and scientists at the only annual event in the world that focuses on mid- to large-scale vermicomposting.

US Composting Council (USCC)

compostingcouncil.org

The USCC is the composting industry's national trade organization. They provide wide-ranging information about composting and compost products. Their annual conference, held the last week in January, brings together new and seasoned composters from throughout the US and other countries. They also have a page that provides information about each state's composting regulations and permitting process: www.composting council.org/state-compost-regulations-map.

BioCycle

www.biocycle.net

BioCycle magazine, biocycle.net, and BioCycle conferences provide a wealth of information to help maximize organics recycling endeavors. The website has additional resources such as "Find A Composter," a directory of equipment and systems, a biweekly bulletin, and an annual report on the state of composting in the United States. They hold three conferences every year and publish scientific articles quarterly in *Compost and Science Utilization*.

Institute for Local Self-Reliance (ILSR)

ilsr.org/composting/

ILSR has a Waste to Wealth program that provides research-based technical assistance to help communities reduce waste and create economic development through composting and recycling. They have webinars, videos, and publications such as *State of Composting in the U.S.: What, Why, Where and How*.

North Carolina Composting Council (NCCC)

carolinacompost.com

The NCCC has been an affiliate association of the USCC since 2007. They promote composting and compost use and hold an annual five-day Compost Operations Training Course.

Vermicomposting Operations Websites

If you'd like to learn more about any of the commercial vermicomposting operations described in this book, the company website is a great place to start.

- **Aldea Verde:** aldeaverde.mx
- **Arizona Worm Farm:** arizonawormfarm.com
- **Big Red Worms:** bigredworms.com
- **California Soils:** californiasoils.com
- **Earthen Organics:** earthenorganics.com
- **Mass Wiggle:** masswiggle.com
- **Noke Ltd.:** mynoke.co.nz
- **PUR VER:** purver.be
- **TerraVesco:** terravesco.com
- **Texas Worm Ranch:** txwormranch.com
- **WormCycle:** wormcycle.com
- **Worm Endings Unlimited:** wormendingsunlimited.com
- **The Worm Farm:** thewormfarm.net
- **The Worm Ladies of Charlestown:** wormladies.com

Notes

Chapter 2: How Vermicast Benefits Soils and Plants

1. R. M. Atiyeh et al., "Changes in biochemical properties of cow manure during processing by earthworms (*Eisenia andrei*, Bouché) and the effects on seedling growth," *Pedobiologia* 44 (2000): 709–24.
2. E. N. Yardim et al., "Suppression of tomato hornworm (*Manduca quinquemaculata*) and cucumber beetles (*Acalymma vittatum* and *Diabotrica undecimpunctata*) populations and damage by vermicomposts," *Pedobiologia* 50 (2006): 23–29.
3. K. R. Rao et al., "Influence of fertilizers and manures on the population of coccinellid beetles and spiders in groundnut ecosystem," *Annals of Plant Protection Sciences* 9 (2001): 43–46; and K. R. Rao, "Induce host plant resistance in the management of sucking pests of groundnut," *Annals of Plant Protection Sciences* 10 (2002): 45–50.
4. M. M. Szczech, "Suppressiveness of vermicomposts against fusarium wilt of tomato," *Journal of Phytopathology* 147 (1999): 155–61.
5. S. Miyasaka, N. Q. Arancon, et al., "Suppressiveness of vermicomposts against fusarium wilt of tomato," Sustainable Agriculture Research & Education, 2015, https://projects.sare.org/project-reports/sw10-013.
6. N. Q. Arancon et al., "Management of plant parasitic nematode populations by use of vermicomposts," in *Proceedings Brighton Crop Protection Conference—Pests and Disease*, 2002, 705–16.
7. N. Q. Arancon et al., "The trophic diversity of nematode communities in soils treated with vermicomposts," *Pedobiologia* 47 (2003): 736–40.

Chapter 3: Key Things to Know About Earthworms

1. Dr. Lee Frelich, personal communication, September 29, 2016.
2. R. N. Carrow, D. V. Waddington, and P. E. Rieke, *Turfgrass Soil Fertility and Chemical Problems: Assessment and Management* (Hoboken, NJ: John Wiley & Sons, 2001), 277.

Bibliography

Addabdo, T. D. "The nematicidal effect of organic amendments: A review of the literature 1982–1994." *Nematologia Mediterranea* 23 (1995): 299–305.

Agarwal, S., R. K. Sinha, J. Sharma, et al. "Vermiculture for sustainable horticulture: Agronomic impact studies of earthworms, cow dung compost and vermi-compost vis-à-vis chemical fertilizers on growth and yield of lady's finger (*Abelmoschus esculentus*)." In "Special Issue on 'Vermiculture Technology,'" *International Journal of Environmental Engineering*. Geneva, Switzerland: Inderscience Publishers, 2010.

Aira, M., F. Monroy, and J. Dominguez. "Earthworms strongly modify microbial biomass and activity triggering enzymatic activities during vermicomposting independently of the application rates of pig slurry." *Science of the Total Environment* 258 (2007): 252–61.

Aira, M., M. Gómez-Brandón, P. González-Porto, and J. Domínguez. "Selective reduction of the pathogenic load of cow manure in an industrial-scale continuous-feeding vermireactor." *Bioresource Technology* 102 (2011): 9633–37.

Akhtar, M., and A. Malik. "Role of organic amendments and soil organisms in the biological control of plant parasitic nematodes: A review." *Bioresource Technology* 74 (2000): 35–47.

Arancon, N. Q., C. A. Edwards, R. Atiyeh, and J. D. Metzger. "Effects of vermicomposts produced from food waste on the growth and yields of greenhouse peppers." *Bioresource Technology* 93 (2004): 139–44.

Arancon, N. Q., C. A. Edwards, A. Babenko, J. Cannon, P. Galvis, and J. D. Metzger. "Influences of vermicomposts, produced by earthworms and microorganisms from cattle manure, food waste and paper waste, on the germination, growth and flowering of petunias in the greenhouse." *Applied Soil Ecology* 39, no. 1 (2008): 91–99.

Arancon, N. Q., C. A. Edwards, P. Bierman, J. D. Metzger, S. Lee, and C. Welch. "Effects of vermicomposts to tomatoes and peppers grown in the field and strawberries under high plastic tunnels." *Pedobiologia* 47 (2003): 731–35.

Arancon, N. Q., C. A. Edwards, P. Bierman, C. Welch, and J. D. Metzger. "The influence of vermicompost applications to strawberries: Part 1. Effects on growth and yield." *Bioresource Technology* 93 (2004): 145–53.

Arancon, N. Q., C. A. Edwards, and S. Lee. "Management of plant parasitic nematode populations by use of vermicomposts." *Proceedings Brighton Crop Protection Conference—Pests and Diseases* (2002): 705–16.

Arancon, N. Q., C. A. Edwards, E. N. Yardim, T. J. Oliver, R. J. Byrne, and G. Keeney. "Suppression of two-spotted spider mite (*Tetranychus urticae*), mealy bug (*Pseudococcus* sp) and aphid (*Myzus persicae*) populations and damage by vermicomposts." *Crop Protection* 26 (2007): 29–39.

Arancon, N. Q., P. A. Galvis, and C. A. Edwards. "Suppression of insect pest populations and damage to plants by vermicomposts." *Bioresource Technology* 96 (2005): 1137–42.

Arancon, N. Q., P. Galvis, C. A. Edwards, and E. Yardim. "The trophic diversity of nematode communities in soils treated with vermicomposts." *Pedobiologia* 47 (2003): 736–40.

Arancon, N. Q., S. Lee, C. A. Edwards, and R. M. Atiyeh. "Effects of humic acids and aqueous extracts derived from cattle, food and paper-waste vermicomposts on growth of greenhouse plants." *Pedobiologia* 47 (2003): 744–81.

Asha, A., A. K. Tripathi, and P. Soni. "Vermicomposting: A better option for organic solid waste management." *Journal of Human Ecology* 24 (2008): 59–64.

Atiyeh, R. M., N. Q. Arancon, C. A. Edwards, and J. D. Metzger. "Influence of earthworm-processed pig manure on the growth and yield of greenhouse tomatoes." *Bioresource Technology* 75 (2000): 175–80.

Atiyeh, R. M., N. Q. Arancon, C. A. Edwards, and J. D. Metzger. "The influence of earthworm-processed pig manure on the growth and productivity of marigolds." *Bioresource Technology* 81 (2001): 103–8.

Atiyeh, R. M., J. Dominguez, S. Subler, and C. A. Edwards. "Changes in biochemical properties of cow manure during processing by earthworms (*Eisenia andrei*, Bouché) and the effects on seedling growth." *Pedobiologia* 44 (2000): 709–24.

Atiyeh, R. M., S. Lee, C. A. Edwards, N. Q. Arancon, and J. D. Metzger. "The influence of humic acids derived from earthworm-processed organic wastes on plant growth." *Bioresource Technology* 84 (2002): 7–14.

Atiyeh, R. M., S. Subler, C. A. Edwards, G. Bachman, J. D. Metzger, and W. Shuster. "Effects of vermicomposts and composts on plant growth in horticulture container media and soil." *Pedobiologia* 44 (2000): 579–90.

Bansal, S. and K. K. Kapoor. "Vermicomposting of crop residues and cattle dung with *Eisenia foetida*." *Bioresource Technology* 73 (2000): 95–98.

Banu, J. R., S. Logakanthi, and G. S. Vijayalakshmi. "Biomanagement of paper mill sludge using an indigenous (*Lampito mauritii*) and two exotic (*Eudrilus eugineae* and *Eisenia foetida*) earthworms." *Journal of Environmental Biology* 22 (2001): 181–85.

Benitez, E., R. Nogales, C. Elvira, G. Masciandaro, and B. Ceccanti. "Enzyme activities as indicators of the stabilization of sewage sludges composting by *Eisenia foetida*." *Bioresource Technology* 67 (1999): 297–303.

Bernal, M. P., C. Faredes, M. A. Sanchez-Monedero, and J. Cegarra. "Maturity and stability parameters of composts prepared with a wide range of organic wastes." *Bioresource Technology* 63 (1998): 91–99.

Bernard, M. D., M. S. Peter, W. D. Christopher, and H. R. Maarten. "Interactions between earthworms, beneficial soil microorganisms and root pathogens." *Applied Soil Ecology* 1 (1994): 3–10.

Binet, F., L. Fayolle, and M. Pussard. "Significance of earthworms in stimulating soil microbial activity." *Biology & Fertility of Soils* 27 (1998): 79–84.

Blok, W. J., J. G. Lamers, A. J. Termoshuizen, and G. J. Bollen. "Control of soil-borne plant pathogens by incorporating fresh organic amendments followed by tarping." *Phytopathology* 90 (2000): 253–59.

Bohlen, P. J., and C. A. Edwards. "Earthworm effects on N dynamics and soil respiration in microcosms receiving organic and inorganic nutrients." *Soil Biology Biochemistry* 27 (1995): 341–48.

Brown, B. A., and M. J. Mitchell. "Role of the earthworm, *Eisenia fetida*, in affecting survival of *Salmonella enteritidis* ser. *Typhimurium*." *Pedobiologia* 22 (1981): 434–38.

Buckerfield, J. C., and K. A. Webster. "Worm-worked waste boosts grape yields: Prospects for vermicompost use in vineyards." *Australian and New Zealand Wine Industry Journal* 13 (1998): 73–76.

Byzov, B. A., T. Y. Nechitaylo, B. K. Bumazhkin, A. V. Kurakov, P. N. Golyshin, and D. G. Zvyagintsev. "Culturable microorganisms from the earthworm digestive tract." *Microbiology* 78 (2009): 360–68.

Canellas, L. P., F. L. Olivares, and F. A. R. Okorokova. "Humic acids isolated from earthworm compost enhance root elongation, lateral root emergence and plasma membrane H+– ATPase activity in maize roots." *Plant Physiology* 130 (2002): 1951–57.

Carrow, R. N., D. V. Waddington, and P. E. Rieke. *Turfgrass Soil Fertility and Chemical Problems: Assessment and Management*. Hoboken, NJ: John Wiley & Sons, 2001.

Chamani, E., D. C. Joyce, and A. Reihanytabar. "Vermicompost effects on the growth and flowering of petunia hybrida 'Dream Neon Rose'." *American-Eurasian Journal of Agriculture and Environmental Science* 3 (2008): 506–12.

Chen, W., H. A. Hoitink, A. F. Schmitthenner, and O. Touvinen. "The role of microbial activity in suppression of damping-off caused by *Pythium ultimum*." *Phytopathology* 78 (1987): 314–22.

Chen, Y., and T. Aviad. "Effects of humic substances on plant growth." In *Humic Substances in Soil and Crop Sciences*, edited by P. R. Bloom, 161–86. Madison, WI: American Society of Agronomy, 1990.

Citernesi, U., R. Neglia, A. Seritti, et al. "Nitrogen fixation in the gastro-enteric cavity of soil animals." *Soil Biology Biochemistry* 9 (1977): 71–72.

Contreras, E. "Studies on the intestinal actinomycete flora of *Eisenia lucens* (Annelida, Oligochaeta)." *Pedobiologia* 20 (1980): 411–16.

Daniel, O., and J. M. Anderson. "Microbial biomass and activity in contrasting soil material after passage through the gut of earthworm *Lumbricus rubellus*." *Soil Biology Biochemistry* 24 (1992): 465–70.

Devliegher, W., and W. Verstraete. "*Lumbricus terrestris* in a soil core experiment: Nutrient-enrichment processes (NEP) and gut-associated processes (GAP) and their effect on microbial biomass and microbial activity." *Soil Biology Biochemistry* 27 (1995): 1573–80.

Dominguez, J., and C. A. Edwards. "Vermicomposting organic wastes: A review." In *Soil Zoology for Sustainable Development in the 21st Century*, edited by S. H. Shakir Hanna and W. Z. A. Mikhail, 369–95. Cairo: self-published, 2004.

Dominguez, J., C. A. Edwards, and M. Webster. "Vermicomposting of sewage sludge: Effects of bulking materials on the growth and reproduction of the earthworm *Eisenia andrei*." *Pedobiologia* 44 (2000): 24–32.

Dominguez, J., J. C. Sanchez-Hernandez, and M. Lores. "Vermicomposting of winemaking by-products." In *Handbook of Grape Processing By-Products*, 55–78. Amsterdam: Elsevier, 2017.

Edwards, C. A. "Breakdown of animal, vegetable and industrial organic wastes by earthworms." In *Earthworms in Waste and Environmental Management*, edited by C. A. Edwards and E. F. Neuhauser, 21–31. Amsterdam: SPB Academic Publishing, 1988.

Edwards, C. A. "Human pathogen reduction during vermicomposting." In *Vermiculture Technology: Earthworms, Organic Wastes and Environmental Management*, edited by C. A. Edwards, N. Q. Arancon, and R. Sherman, 249–61. Boca Raton, FL: CRC Press, 2011.

Edwards, C. A., N. Q. Arancon, M. V. Bennett, A. Askar, and G. Keeney. "Effect of aqueous extracts from vermicomposts on attacks by cucumber beetles (*Acalymna vittatum*) (Fabr.) on cucumbers and tobacco hornworm (*Manduca sexta*) (L.) on tomatoes." *Pedobiologia* 53 (2010): 141–48.

Edwards, C. A., N. Q. Arancon, M. V. Bennett, A. Askar, G. Keeney, and B. Little. "Suppression of green peach aphid (*Myzus persicae*) (Sulz.), citrus mealybug (*Planococcus citri*) (Risso), and two spotted spider mite

(*Tetranychus urticae*) (Koch.) attacks on tomatoes and cucumbers by aqueous extracts from vermicomposts." *Crop Protection* 29 (2010): 80–93.

Edwards, C. A., and P. J. Bohlen. *Biology and Ecology of Earthworms*. London: Chapman and Hall, 1996.

Edwards, C. A., and I. Burrows. "The potential of earthworm composts as plant growth media." In *Earthworms in Waste and Environmental Management*, edited by C. A. Edwards and E. Neuhauser, 21–32. Amsterdam: SPB Academic Press, 1988.

Edwards, C. A., J. Dominguez, and N. Q. Arancon. "The influence of vermicomposts on pest and diseases." In *Soil Zoology for Sustainable Development in the 21st Century*, edited by S. H. Shakir Hanna and W. Z. A. Mikhail, 397–418. Cairo: self-published, 2004.

Edwards, C. A., and R. Lofty. *The Biology of Earthworms*. London: Chapmann and Hall, 1977.

Garg, P., A. Gupta, and S. Satya. "Vermicomposting of different types of waste using *Eisenia foetida*: A comparative study." *Bioresource Technology* 97 (2006): 391–95.

Ghosh, M., G. N. Chattopadhyay, and K. Baral. "Transformation of phosphorus during vermicomposting." *Bioresource Technology* 69 (1999): 149–54.

Gunadi, B., C. Blount, and C. A. Edwards. "The growth and fecundity of *Eisenia foetida* (Savigny) in cattle solids pre-composted for different periods." *Pedobiologia* 46 (2002): 15–23.

Hahn, H., and M. Bopp. "A cytokinin test with high specificity." *Planta* 83 (1968): 115–18.

Haritha, Devi S., K. Vijayalakshmi, K. Pavana Jyotsna, et al. "Comparative assessment in enzyme activities and microbial populations during normal and vermicomposting." *Journal of Environmental Biology* 30 (2009): 1013–17.

Hartenstein, R., E. F. Neuhauser, and D. L. Kaplan. "Reproductive potential of the earthworm *Eisenia foetida*." *Oecologia* 43 (1979): 329–40.

Kale, R. D., and N. Karmegam. "The role of earthworms in tropics with emphasis on Indian ecosystems." *Applied and Environmental Soil Science* (2010): Article ID 414356.

Kannangowa, T., R. S. Utkhede, J. W. Paul, and Z. K. Punja. "Effect of mesophilic and thermophilic composts on suppression of *Fusarium* root and stem rot of greenhouse cucumber." *Canadian Journal of Microbiology* 46 (2000): 1021–22.

Karmegam, N., and T. Daniel. "Effect of biodigested slurry and vermicompost on the growth and yield of cowpea (*Vigna unguiculata* (L.)." *Environmental Ecology* 18 (2000): 367–70.

Kostecka, J., J. B. Blazej, and M. Kolodziej. "Investigations on application of vermicompost in potatoes farming in second year of experiment." *Zeszyty Naukowe Akademii Rolniczej W Krakowie* 310 (1996a): 69–77.

Krishnamoorthy, R. V., and S. N. Vajranabhiah. "Biological activity of earthworm casts: An assessment of plant growth promoter levels in casts." *Proceedings of Indian Academy of Science (Animal Science)* 95 (1986): 341–45.

Kristufek, V., K. Ravasz, and V. Pizl. "Actinomycete communities in earthworm guts and surrounding soil." *Pedobiologia* 37 (1993): 379–84.

Lazarovits, G., M. Tenuta, K. L. Conn, et al. "Utilization of high nitrogen and swine manure amendments for control of soil-borne diseases: Efficacy and mode of action." *Acta Horticulturae* 5 (2000): 559–64.

Lazcano, C., M. Gomez-Brandon, and J. Dominguez. "Comparison of the effectiveness of composting and vermicomposting for the biological stabilization of cattle manure." *Chemosphere* 72 (2008): 1013–19.

Lee, Y. S., and R. J. Bartlett. "Stimulation of plant growth by humic substances." *Soil Science Society of America Journal* 40 (1976): 876–79.

Manivannan, S., M. Balamurugan, K. Parthasarathi, et al. "Effect of vermicompost on soil fertility and crop productivity—beans (*Phaseolus vulgaris*)." *Journal of Environmental Biology* 30 (2009): 275–81.

Mather, U., L. K. Verma, J. N. Srivastava. "Effects of vermicomposting on microbiological flora of infected biomedical waste." *ISHWM Journal* 5 (2006): 28–33.

Mitchell, A., and C. A. Edwards. "The production of vermicompost using *Eisenia fetida* from cattle manure." *Soil Biology Biochemistry* 29 (1997): 3–4.

Mitchell, M. J., S. G. Hornor, and B. I. Abrams. "Decomposition of sewage sludge in drying beds and the potential role of the earthworm, *Eisenia foetida*." *Journal of Environmental Quality* 9 (1980): 373–78.

Miyasaka, S., N. Q. Arancon, L. Cox, et al. "Control of bacterial wilt disease of ginger through an integrated pest management program." Final Report for SW10-013, Sustainable Agriculture Research & Education. 2015. https://projects.sare.org/project-reports/sw10-013.

Monroy, F., M. Aira, and J. Domínguez. "Reduction of total coliform numbers during vermicomposting is caused by short-term direct effects of earthworms on microorganisms and depends on the dose of application of pig slurry." *Science of the Total Environment* 407 (2009): 5411–16.

Muscolo, A., F. Bovalo, F. Gionfriddo, and S. Nardi. "Earthworm humic matter produces auxin-like effect on *Daucus carota* cell growth and nitrate metabolism." *Soil Biology Biochemistry* 31 (1999): 1303–11.

Muscolo, A., M. Felici, G. Concheri, and S. Nardi. "Effect of earthworm humic substances on esterase and peroxidase activity during growth of leaf explants of *Nicotiana plumbaginifolia*." *Biology and Fertility of Soils* 15 (1993): 127–31.

Muscolo, A., M. R. Panuccio, M. R. Abenavoli, et al. "Effect of molecular complexity and acidity of earthworm faeces humic fractions on glutamate dehydrogenase, glutamine synthetase, and phosphoenolpyruvate carboxylase in *Daucus carota* II cells." *Biology and Fertility of Soils* 22 (1996): 83–88.

Nakamura, Y. "Interactions between earthworms and microorganisms in biological control of plant root pathogens." *Farming Japan* 30 (1996): 37–43.

Ndegwa, P. M., S. A. Thompson, and K. C. Das. "Effects of stocking density and feeding rate on vermicomposting of biosolids." *Bioresource Technology* 71 (2000): 5–12.

Nielson, R. L. "Presence of plant growth substances in earthworms demonstrated by paper chromatography and the Went pea test." *Nature* 208 (1965): 1113–14.

Orozco, F. H., J. Cegarra, L. M. Trujillo, and A. Roig. "Vermicomposting of coffee pulp using the earthworm *Eisenia fetida*: Effects on C and N contents and the availability of nutrients." *Biology and Fertility of Soils* 22 (1996): 162–66.

Parthasarathi, K., and L. S. Ranganathan. "Pressmud vermicast are hot spots of fungi and bacteria." *Ecology and Environmental Conservation* 4 (1998): 81–86.

Pathma, J., R. Kamaraj Kennedy, and N. Sakthivel. "Mechanisms of fluorescent pseudomonads that mediate biological control of phytopathogens and plant growth promotion of crop plants." In *Bacteria in Agrobiology: Plant Growth Responses*, edited by D. K. Maheswari, 77–105. Berlin: SpringerVerlag, 2011.

Pathma, J., and N. Sakthivel. "Microbial diversity of vermicompost bacteria that exhibit useful agricultural traits and waste management potential." *Springerplus* 1 (2012): 26.

Radovich, Theodore, and Norman Arancon, eds. *Tea Time in the Tropics: A Handbook for Compost Tea Production and Use*. Honolulu: College of Tropical Agriculture and Human Resources, University of Hawaii, 2011. https://www.sare.org/Learning-Center/SARE-Project-Products /Western-SARE-Project-Products/Tea-Time-in-the-Tropics.

Rao, K. R. "Induce host plant resistance in the management of sucking pests of groundnut." *Annals of Plant Protection Sciences* 10 (2002): 45–50.

Rao, K. R., P. A. Rao, and K. T. Rao. "Influence of fertilizers and manures on the population of coccinellid beetles and spiders in groundnut ecosystem." *Annals of Plant Protection Sciences* 9 (2001): 43–46.

Rivera, A. M. C., E. R. Wright, M. V. López, and M. C. Fabrizio. "Temperature and dosage dependent suppression of damping-off caused by *Rhizoctonia solani* in vermicompost amended nurseries of white pumpkin." *Phyton International Journal of Experimental Botany* 53 (2004): 131–36.

Rodriguez, J. A., E. Zavaleta, P. Sanchez, and H. Gonzalez. "The effect of vermicomposts on plant nutrition, yield and incidence of root and crown rot of gerbera (*Gerbera jamesonii* H. Bolus)." *Fitopatol* 35 (2000): 66–79.

Rodriguez-Kabana, R. "Organic and inorganic amendments to soil as nematode suppressants." *Journal of Nematology* 18 (1986): 129–35.

Sánchez-Monedero, M. A., A. Roig, C. Paredes, and M. P. Bernal. "Nitrogen transformation during organic waste composting by the Rutgers system and its effects on pH, EC and maturity of the composting mixtures." *Bioresource Technology* 78 (2001): 301–8.

Scheu, S. "Automated measurement of the respiratory response of soil micro-compartments: Active microbial biomass in earthworm faeces." *Soil Biology and Biochemistry* 24 (1992): 1113–18.

Scheuerell, S. J., D. M. Sullivan, and W. F. Mahaffee. "Suppression of seedling damping-off caused by *Pythium ultimum* and *Rhizoctonia solani* in container media amended with a diverse range of Pacific Northwest compost sources." *Phytopathology* 95 (2005): 306–15.

Schmidt, O., B. M. Doubre, M. H. Ryder, and K. Killman. "Population dynamics of *Pseudomonas corrugata* 2140R LUX8 in earthworm food and in earthworm cast." *Soil Biology and Biochemistry* 29 (1997): 523–28.

Senesi, N., C. Saiz-Jimenez, and T. M. Miano. "Spectroscopic characterization of metal-humic acid-like complexes of earthworm-composted organic wastes." *Science of the Total Environment* 117–118 (1992): 111–20.

Simsek Ersahin, Y., K. Haktanir, and Y. Yanar. "Vermicompost suppresses *Rhizoctonia solani* Kühn in cucumber seedlings." *Journal of Plant Disease Protection* 9 (2009): 15–17.

Singh, R., R. R. Sharma, S. Kumar, R. K. Gupta, and R. T. Patil. "Vermicompost substitution influences growth, physiological disorders, fruit yield and quality of strawberry (*Fragaria × ananassa* Duch.)." *Bioresource Technology* 99 (2008): 8507–11.

Singh, U. P., S. Maurya, and D. P. Singh. "Antifungal activity and induced resistance in pea by aqueous extract of vermicompost and for control of powdery mildew of pea and balsam." *Journal of Plant Disease Protection* 110 (2003): 544–53.

Sipes, B. S., A. S. Arakaki, D. P. Schmitt, and R. T. Hamasaki. "Root-knot nematode management in tropical cropping systems with organic products." *Journal of Sustainable Agriculture* 15 (1999): 69–76.

Szczech, M. "Suppressiveness of vermicomposts against fusarium wilt of tomato." *Journal of Phytopathology* 147 (1999): 155–61.

Szczech, M., W. Rondomanski, M. W. Brzeski, U. Smolinska, and J. F. Kotowski. "Suppressive effect of a commercial earthworm compost on some root infecting pathogens of cabbage and tomato." *Biology of Agriculture and Horticulture* 10 (1993): 47–52.

Szczech, M., and U. Smolinska. "Comparison of suppressiveness of vermicomposts produced from animal manures and sewage sludge against *Phytophthora nicotianae* Breda de Haan var. *nicotiannae*." *Journal of Phytopathology* 149 (2001): 77–82.

Tajbakhsh, J., M. A. Abdoli, E. Mohammadi Goltapeh, et al. "Trend of physico-chemical properties change in recycling spent mushroom compost through vermicomposting by epigeic earthworms *Eisenia foetida* and *E. andrei*." *Journal of Agricultural Technology* 4 (2008): 185–98.

Thoden, T. C., G. W. Korthals, and A. J. Termorshuizen. "Organic amendments and their influences on plant-parasitic and free living nematodes: A promising method for nematode management." *Nematology* 13 (2011): 133–53.

Tiquia, S. M. "Microbiological parameters as indicators of compost maturity." *Journal of Applied Microbiology* 99 (2005): 816–28.

Tomati, U., A. Grappelli, and E. Galli. "The presence of growth regulators in earthworm worked waste." In *On Earthworms. Proceedings of International Symposium on Earthworms, Selected Symposia and Monographs, Union Zoologica Italian, 2*, edited by A. M. Bonvicini Paglioi and P. Omodeo, 423–35. Modena, Italy: Mucchi, 1987.

Toyota, K., and M. Kimura. "Microbial community indigenous to the earthworm *Eisenia foetida*." *Biology and Fertility of Soils* 31 (2000): 187–90.

Tringovska, I., and T. Dintcheva. "Vermicompost as substrate amendment for tomato transplant production." *Sustainable Agriculture Research* 1, no. 2 (2012): 115.

Valdrighi, M. M., A. Pera, M. Agnolucci, et al. "Effects of compost-derived humic acids on vegetable biomass production and microbial growth within a plant (*Cichorium intybus*) soil system: A comparative study." *Agriculture, Ecosystems & Environment* 58 (1996): 133–44.

Vinken, R., A. Schaeffer, and R. Ji. "Abiotic association of soil-borne monomeric phenols with humic acids." *Organic Geochemistry* 36 (2005): 583–93.

Vivas, A., B. Moreno, S. Garcia-Rodriguez, and E. Benitez. "Assessing the impact of composting and vermicomposting on bacterial community size and structure, and functional diversity of an olive-mill waste." *Bioresource Technology* 100 (2009): 1319–26.

Webster, K. A. "Vermicompost increases yield of cherries for three years after a single application." *South Australia: EcoResearch* (2005).

Yardim, E. N., N. Q. Arancon, C. A. Edwards, et al. "Suppression of tomato hornworm (*Manduca quinquemaculata*) and cucumber beetles (*Acalymma vittatum* and *Diabotrica undecimpunctata*) populations and damage by vermicomposts." *Pedobiologia* 50 (2006): 23–29.

Yardim, E. N., and C. A. Edwards. "Effects of organic and synthetic fertilizer sources on pest and predatory insects associated with tomatoes." *Phytoparasitica* 31 (2003): 324–29.

Yasir, M., Z. Aslam, S. W. Kim, et al. "Bacterial community composition and chitinase gene diversity of vermicompost with antifungal activity." *Bioresource Technology* 100 (2009): 4396–403.

Index

Note: Page numbers in italics refer to figures and photos. Page numbers followed by t refer to tables.

About the Author

Extension Specialist Rhonda Sherman is the director of the Compost Learning Lab at North Carolina State University and one of the leading experts worldwide on vermicomposting. Rhonda travels extensively to present workshops and to consult with farmers, businesses, and institutions on the development and management of vermicomposting systems. She organizes the highly successful annual NC State Vermiculture Conference, which for 19 years has drawn participants from across the United States and around the globe. She is a coeditor of *Vermiculture Technology* and has written extensively about composting and vermicomposting in her role with NC State University. Sherman is also a cofounder of the North Carolina Composting Council (NCCC) and a past board member of the US Composting Council, and she has served as a contributing editor to *BioCycle Journal of Composting & Organics Recycling*.

Natalie Hampton